# Moedim:
## Celestial Events in The Heavens

## What is God Trying to Tell Us?

**By: Miles R. Jones, Ph.D.**

**Chapter 13** *"Marrage At Sinai"* **Written by:**
**Catherine Stewart-Jones**


ISBN # 978-1-957488-08-0

*"Yehovah gave the Word; Great was the company of those that published it!" Psalm 68:11*

**Great Publishing Company**
Benai Emunah Institute
Kerrville, Texas

# About the Book

Dr. Jones' groundbreaking work establishing the ancient (biblical) timeline is gathered together here in one volume. Taken from his books ***Writing of God*** and ***Sons of Zion*** - Dr. Jones adds even more information on celestial events and the 2023 and 2024 eclipses intersecting in Texas. The all-important biblical chronology has been fixed precisely by the retro-calculation of celestial events recounted in the ancient records. Using this scientific methodology - called the "New Chronology "by the archaeologist David Rohl - exact dates for these ancient celestial events, like eclipses, can be synchronized with the Bible story and the records of ancient civilizations, their kings, monuments and artifacts. The anti-God proponents in the academic system have long used the story of the Exodus to prove the Bible is an inaccurate history - thereby claiming the message of the Bible must be inaccurate as well. **Dr. Jones has put the Bible back into the history business - proving the truth of the Bible story with amazing scientific exactitude!**

# About the Author

Dr. Miles R. Jones is a biblical archaeologist with three degrees in languages and linguistics. He created the Benai Emunah Institute to locate, translate and publish ancient biblical inscriptions such as those found at the base of Mount Sinai in Midian, Arabia. The Institute has recovered dozens of Hebrew manuscripts of the New Testament and authenticated some as coming from first century sources. The publication of the Hebrew Book of Revelation is one example.

His previous work as a director of university English programs in the United Arab Emirates and a specialist with the U.S. government in Yemen, training military officers, allowed him to begin his research in the field. As an historical linguist, the origin of the alphabet had always captivated him. That original alphabet appeared in the path of the Exodus at the time of the Exodus. In his travels Dr. Jones continually collected and analyzed the data. This book is the culmination of decades of research. Dr. Jones received a PhD in Foreign Languages at the University of Texas, at Austin in 1985. For more information go to **writingofgod.com**.

# Contents

Moedim book layout and design by Patrick J. McGuire, Altarnet Media Ministries

# Preface

By Dr. Miles R. Jones

**Before we discuss the meaning of celestial events, we must first know what they are.** Solar and lunar eclipses would come first to our minds but there is much more. Solar cycles determine the seasons and the feast days of Yehovah. Yeshua the Messiah (Jesus Christ in Greek) was crucified on Passover and is said to have been born on Sukkot (Feast of Tabernacles) and on that feast day of Sukkot he shall return, if the sages are correct.

Lunar monthly cycles determined the days of gathering and worship each month. The lunar cycle is from new moon to full moon, a time span that is irregular but still determined by the orbit of the moon. It is an average of 29.53 days - sometimes it is 29 days and sometimes 30 days - because of this irregularity it can be retro-calculated. An ancient record of the lunar cycles can tell us the exact date of them even thousands of years later.

Related to the solar and lunar cycles are the Venus cycles. Venus crosses between the earth and the sun, seemingly irregular but the Venus orbit can be calculated for the past and the future. The cycle has Venus in close approach twice in eight years, the last time was 2004 and 2012, then again 105 years later, that will be 2117, then another 121 years later in 2238, for a total of the cycle which repeats every 243 years. This can be helpful in fixing the ancient timeline when recorded regularly in the annals.

Also known are the cycles of the comets. Hailey's comet comes around about every 75-76 years. It is said that Samuel Clemens - also known as Mark Twain - was born under the comet and died under it almost 75 years later. Let us not forget asteroids that fall to the earth in swarms lighting up the night sky in beauty - or more rarely a large one impacts in a huge catastrophic event.

There are also the cycles of the night sky which are designated by the astronomical patterns of the stars - the signs of the Zodiac. They rotate slowly across the sky in a 24-hour cycle, and they change by season and location. Although much misused by astrologists - they are still celestial events of God and many verses in Scripture that seem so cryptic are referring to those signs, like the woman with the 12 stars in Revelation 12:6.

There are also celestial events in the lower heavens, such as blood moons which are caused by atmospheric effects during a lunar eclipse. **The phenomenon of four bloods moons on Passover and Sukkot two years in row is called a Tetrad.** They are rare but the Tetrad has occurred three times in the past 75 years. They foretell difficult times for the Hebrews.

The nation of Israel was reborn May 14, 1948 and was immediately pitched into battle during the Tetrad of 1949-50 with the Arabs who did not welcome this event. During the Tetrad of 1967-68, Israel fought against Egypt, Jordan and Syria in the *6-day War* with the result that Israel regained Jerusalem. During the Tetrad of 2014-15, Israel defended against Hamas in what was called the *Tunnel War*. Tunnels were dug beneath the border to launch a surprise attack. These were all fateful events for Israel.

Before these three Tetrads within the past century there had not been any for almost 500 years. In 1492, King Ferdinand and Queen Isabella of Spain completed the *Reconquista* by taking back the last stronghold of the Moors at Granada. They immediately proceeded to expel the Jews from Spain forcing them to convert or leave by August of 1492. On that deadline, Christopher Columbus left Spain for the new world with a ship full of Messianics fleeing the Inquisition, and Jews fleeing the Expulsion.

One thing that left Spain with the Messianics fleeing the Inquisition was a manuscript of *The Hebrew Gospels* which I recovered from the Vatican Library in 2014-15 during the last Tetrad. **I translated it and was able to conduct services from it for the first time in almost 500 years since the Tetrad in 149394. It was quite a moving experience for me!**

Extraordinary weather events used as celestial signs should never be ignored - like the hailstorm that destroyed the Canaanite army coming against the Israelites in Joshua 10:11, or the fearful storm that Yeshua calmed in Matthew 8:23, or Yehovah speaking to the Israelites from a dark cloud over Mount Sinai with thundering and lightning. During our expedition to Mount Sinai in Arabia - March 2023 - there were miraculous weather events similar to Exodus 19. Yeshua himself said in Luke 21:11 *"Great earthquakes shall be in diverse places, and pestilences, and fearful sights, and great signs shall there be from heaven!"* When Yehovah is trying to get our attention - we should listen!

All of these celestial events have been used by the Creator of the Universe to communicate to His people. God continues to do so.

# What is Yehovah trying to tell us now?

As I write these words I am awaiting the total eclipse of April 8, 2024. I am in Kerrville, Texas, where we just had another eclipse on October 14, 2023 - less than six months ago. Eclipses happen in any one place on the average of every 360 years. We have two within six months of each other! If God is trying to get our attention - it is working!

The last one was an annular eclipse, the moon totally centered in the sun but too small to blot all the sun out. It is called a *"Ring of Fire"* eclipse because the sun is still visible around the moon. This coming eclipse is a total eclipse. It is also quite personal to me since the line of sight of the eclipse runs right across the property of the Benai Emunah Institute which I direct. **I just cannot escape the conclusion that God is trying to get my personal attention!**

The eclipse in Nineveh, Assyria in 763 BC was clearly recorded in their records. In Israel it was during the reign of Jeroboam. At that time, there was a prophet named Jonah. Yehovah sent Jonah to Nineveh to warn them to repent or be destroyed because the once great city had sunk deep into iniquity. Jonah was very reluctant because they were not his people. He finally arrived in Nineveh and preached his message. There was an eclipse. There were riots. There were plagues. The king and the people repented and Yehovah did not destroy the city! It was a warning - the people took heed and were saved - a hopeful lesson.

What does that mean to us that the eclipse of Nineveh in 763 BC was a warning from God that prevented their destruction? **The eclipse here in Kerrville appeared April 8, 2024 at 12:14 pm. Did you know there is a town named Nineveh in Texas?**

The eclipse appeared over **Nineveh, Texas** at 12:22 pm

The eclipse appeared over **Nineveh, Missouri** at 1:59 pm

The eclipse appeared over **Nineveh, Indiana** at 3:05 pm

The eclipse appeared over **Nineveh, Ohio** at 3:08 pm

The eclipse appeared over **Nineveh, Pennsylvania** at 3:15 pm

The eclipse appeared over **Nineveh, Virginia** at 3:20 pm

The eclipse appeared over **Nineveh, New York** at 3:24 pm

The eclipse appeared over **Nineveh, Nova Scotia** at 4:39 pm

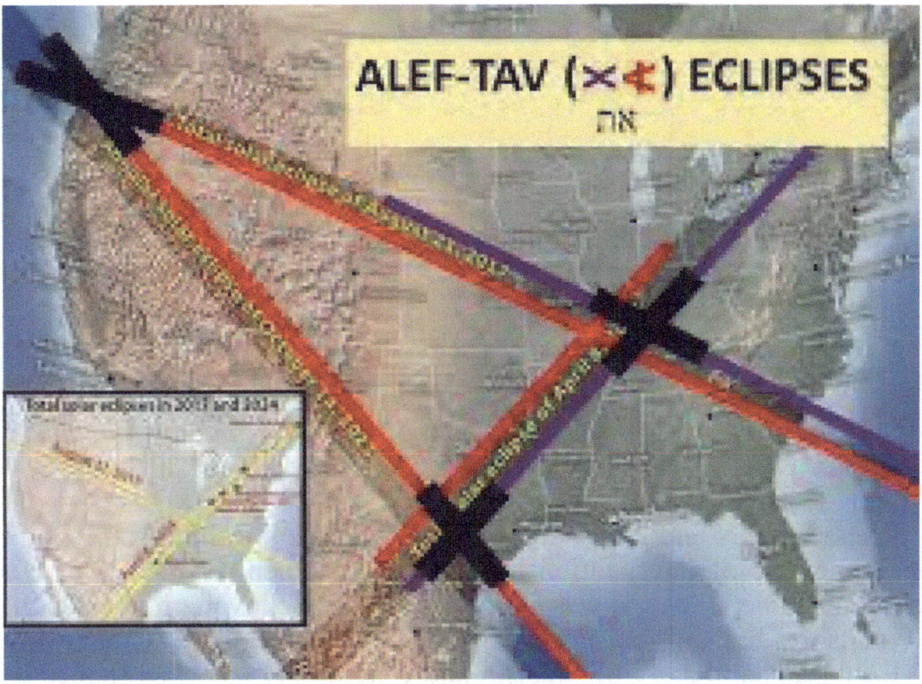


# Let us take heed of God's warning - repent and be saved!

# Chapter One

## Moedim: The Appointed Times

*Let there be lights in the firmament of the heaven to divide the day from the night, and let them be for signs, and for Moedim [Appointed Times]* (Genesis 1:14)

The Exodus and the giving of the law at Mount Sinai were God's supreme acts of intervention in the Old Testament on behalf of His chosen people. Genesis was the prelude when Yehovah prophesied the Exodus to Abraham four hundred years before it occurred (Gen 15:16). All subsequent history of Judaism and Christianity has been shaped by the Exodus and the events of Sinai. There, Moses first prophesied the coming of the Messiah, "*And Yehovah said unto me... I will raise them up a Prophet from among their brethren, like unto you* [Moses], *and will put My words in his mouth; and he shall speak unto them all that I shall command him.*" (Deut 18:15-22). In turn, Yeshua [*Jesus* - in Greek] staked his authority on the Torah received from Har Yehovah [Mountain of God], "*Think not I am come to destroy the Torah, or the prophets; I am not come to destroy but to fulfill.*" (Mt 5:17-18).

Over time the precepts coming from the Sinai Covenant have formed the foundation of Judeo-Christian religion, morality, human rights, and legal justice. The call from Sinai is being debated today with a relevance as crucial as the day Moses descended from Har Yehovah with the tablets containing the Word and the writing of God (Ex 32:15-16).

To dismiss the events at Har Yehovah is to dismiss the foundation of the moral authority of God, if not His very existence! If the foundation is false the entire edifice of religion is a house built upon sand. God becomes a myth. The Gospels - in any language - become irrelevant! Believers are surprised to discover that authorities do not even know the date of the Exodus nor the true location of the Mountain of God.

**The key of knowledge is the word "*Moedim*" above, which means "*appointed times*" signaled to us by the lights of celestial events in the firmament of heaven.**

## The Wrong Time & The Wrong Place

Regarding the date of the Exodus and Sinai Covenant - historians are looking in the wrong time as well as the wrong place. **So how do experts arrive at the dates so confidently, and quite often erroneously, displayed in documentaries, books and museums of ancient history?** In ancient times the date of an event was designated by the year of the reign of the current king. Here is a biblical example:

*And it came to pass in the four hundred and eightieth year after the children of Israel were come out of the land of Egypt, in the fourth year of Solomon's reign over Israel, in the month Zif, which is the second month, that he began to build the house of Yehovah.* (1 Kings 6:1)

A brief description should give you an idea of how all historians approach the historical puzzle. **One has to begin with historical dates that are certain. We call these dates - *anchors*.** The earliest date we are absolutely certain about is the sacking of Thebes in upper Egypt in 664 BC by the invading Assyrians in the 4th year of the reign of King Ashurbanipal, during the 26th (and last) year of the reign of Pharaoh Taharka in Egypt. That leaves three millennia of previous historical dates uncertain!

**The linking of two events, especially the reigns of two major kings in the Mid-East, is called a *synchronism*.** This particular synchronism allows us to correlate, or lock together, the timeline of Egypt with the timeline of Assyria at that moment in ancient history. In the case of the sacking of Thebes we have Ashurbanipal to thank for this synchronism. Not only was Ashurbanipal a great military leader but an outstanding scholar as well. He created a huge library at Nineveh centuries before the Library of Alexandria. Unlike the Alexandrian Library, a large part of Ashurbanipal's library survived. Later invasions left the library in ruin and neglect and it was eventually covered over. Cuneiform tablets, however, are hard to destroy if left intact. In this case, some 1200 manuscripts (on 20,000 tablets) were unearthed, including the genealogy of all the Assyrian kings from time immemorial almost complete.

# We call this a *king list*.

Both the Assyrian and the Babylonian chronicles confirm the sacking of Thebes in 664 BC - along with the ancient Egyptian historian Manetho who lived in the 3rd century BC in Alexandria and had their great library as a resource. We would know so much more about ancient history if the Alexandrian Library had survived. Also in the 3rd century BC, the Chaldean priest Berosus wrote of the event in Greek - citing from Hammurabi's library in Babylon. Surviving monuments confirm the reign of these kings as do the analyses of archaeological findings and the record of astronomical sightings.

**The orbit of the stars and planets can be calculated by modern mathematical astronomers and celestial events - such as eclipses - and the orbits of comets and planets, such as Venus, can be predicted for the future and retro-calculated for the past**. For example, Jonah's eclipse, called the Bur-Sagale, named after the ancient official who recorded it, was visible from Assyria in 763 BC a century before the sacking of Thebes in 664 BC. This allows us to date the Assyrian king list from that we have from Ashurbanipal's library for that epoch.

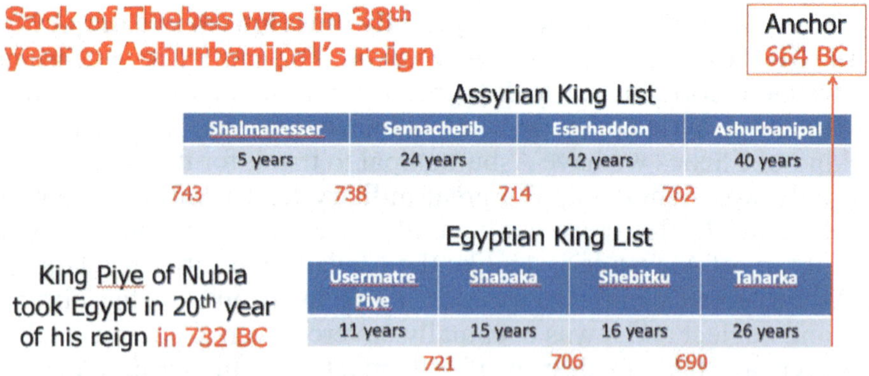


The Egyptians had an elaborate record of keeping Apis bulls which represented the god of the Apis cult. The Apis bull was sacrificed every

25 years so that it would always be in its prime. Another Apis bull was then found to have the proper portents as the reincarnation of the Apis bull god. The sacrificed bull would be interred in a tomb fit for a pharaoh and the celebration throughout Egypt for the new Apis bull god made many a pharaoh groan at the extravagant expense of the ritual. In any case, like clockwork every 25 years the Apis bull was entombed and the regnal year of the pharaoh inscribed. We have those records.

**Edwin Thiele (1983) was the mastermind who scientifically validated the biblical chronology.** Let us follow Thiele's logic. We can synchronize the Bur-Sagale eclipse via the Assyrian Eponym Canon which recorded the eclipse. This Canon was part of the surviving record of Ashurbanipal's library. When an official was named to keep the astronomical history, he was called an Eponym and notated in the records by his name and the year of the reign of the current king. So, strangely enough, lists of bureaucrats and bulls are some of our best witnesses to the reigns of the Assyrian and Egyptian kings of ancient history.

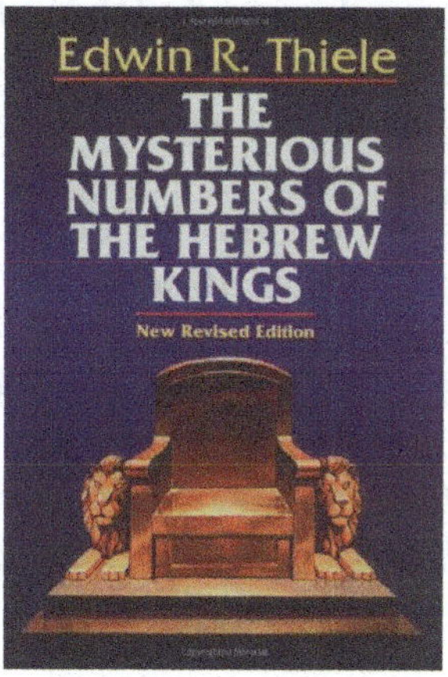


# Chronology is the Backbone of History

Thiele took the astronomical retro-calculation of the Bur-Sagale eclipse on June 15, 763 BC, which is recorded in the Assyrian Eponym Canon in the 9th year of King Ashur-Dan III. Counting back in the Assyrian king list to his ancestor Shalmaneser III gives us a date for his reign from 858-824 BC. This was the king who marched into Canaan in the 6th year of his reign to fight a confederacy of kings, one of which was King Ahab of Israel (1st Kings 22), husband of Jezebel, who sent 2000 chariots and 10,000 warriors into battle against Shalmaneser. In this way, Thiele anchored a scientifically fixed date for the biblical chronology. Thiele was not the first to calculate this timeline from Scripture. It had been done by Bishop James Ussher (1658), Sir Isaac Newton (1728), and more recently Floyd Jones (1993) among others. **Thiele was the first to synchronize the Bible timeline via ancient astronomical records and modern retro-calculation matching the dates of the biblical chronology.**

We can now calculate Ahab's reign from 874-853 BC, and from sources outside the Bible fix the last year of his reign when he was killed at the battle of Karkar. Now we count back in the biblical chronology of kings of the divided monarchy of Judah and Israel. This is no small feat because the two kingdoms used different means of reckoning the calendar. This brings us to 970 BC as the year Solomon ascended the throne for his 40-year reign and therefore 930 BC, the year of his death, as the date of the "*schism*" when the country was divided into a southern kingdom of Judah under the rule of Solomon's son Rehoboam, and the northern kingdom of Israel where a former top official of Solomon, Jeroboam, assumed power.

Continuing backward in time from the dates given in 1st Kings 6:1, 480 years after the Exodus Solomon began building the Temple in the 4th year of his reign. This gives us a date of 1446 BC for the Exodus from Egypt. This is confirmed by the dates given in Judges 11:26 where Jephthah (*circa* 1100 BC) states the Israelites began the conquest of Canaan 300 years previously which would give us 1400 BC. Now we have a date for the onset of the 40 years of wandering and the conquest of Canaan which followed the Exodus.

## The Orthodox Chronology

Despite Thiele's excellent methodology, mainstream academics went in a different direction and arrived at a completely different conclusion from the biblical chronology. Ironically enough they based this new analysis on a passage from the Bible. "*Now there arose up a new pharaoh in Egypt who did not know Joseph... Therefore, they did set over them taskmasters to afflict* [the children of Israel] *with their burdens. And they built for Pharaoh treasure cities, Pi'Thom and Raam'ses*" (Exodus 1:811). The Bible does not cite by name the pharaoh of the Exodus. However, given this passage, **it is assumed the Hebrews built the royal city of Ramses II (the Great),** known as Pi-Ramesse meaning "*the estate of Ramesses,*" on the Pelusiac branch of the Nile. **In the conventional chronology, Ramses II ascended the throne in 1279 BC, therefore secularists**

**insist the Exodus must have happened in the 13th century BC, two centuries later than the biblical chronology. This is called the**

## "*Ramesses Exodus Theory.*"

> Between the 1930s and 1950s evidence accumulated in favor of dating the Exodus and Conquest to the 13th century BC. **That date has remained the majority view.** Even some of those scholars who reject the historicity of the Exodus and Conquest traditions still look to the decades around 1200 BC as the time when Israel emerged... in Canaan. **I will argue here that retention of the 13th century date is an example of scholarly inertia - and the evidence in its favor has long since been eroded away**. (John Bimson, 1988)

So how do we deal with this differing mainstream viewpoint - after examining the precision of Thiele's analysis of the biblical chronology? Did Thiele get something wrong or is something wrong with the Ramesses Exodus Theory? Let us start with the passage in Exodus 1:11 when the Hebrews are building "*Pi-Thom and Raamses*" under the oppression of a new pharaoh who "*knew not Joseph.*" First, note the Hebrew words say the "*estates of Thom and Raamses*" since these two pharaohs did not rule at the same time, Thom's city was built first and later that of Raamses. So, Thom was also a pharaoh before the Exodus, before Ramesses. The Bible says despite the oppression the children of Israel "*multiplied and grew.*" In other words, generations passed.

## Ramesses Was Not the Pharaoh of the Exodus

The Egyptians instituted a program of killing Hebrew babies. A second time Scripture says "*the people multiplied, and waxed very mighty*" (Ex 1: 12 & 20), more generations passed. Then Moses was born, grew to manhood, was exiled and returned at the age of 80 years old to liberate the Israelites from bondage. From Raamses to the Exodus is a span of about two centuries.

If the Israelites were toiling on his royal city about two centuries before the Exodus - **how could Ramesses (aka Raamses) possibly be the pharaoh of the Exodus?** The biblical Exodus was during the 15th century BC, but the Ramesses Exodus Theory places the Conquest in the

13th century, "*searching for evidence that Israel's conquest of Canaan occurred at the close of the... 13th century BC, scholars have failed to find any convincing correlations*" (Bimson, 1988). Therefore, "*it is no longer possible to offer even a reasonable defense of the Conquest narratives*" (Lemche, 1985). **No Conquest in the 13th century BC means, to secularists, that the Bible is myth not real history.** "The *Exodus is a history that never happened*" (Herzog, 1999) Secular theorists declared Ramesses the pharaoh of the Exodus, then declared the Bible false due to a lack of archaeological evidence based on that Orthodox Timeline. It is a tautology, a circular argument, a logical fallacy caused by a false premise.

The misinterpretation of Exodus 1:11 turns on the name of the city of Raamses. David Rohl untangled this mess by pointing out that the **reference to the royal city of Raamses is an *anachronism*.** An *anachronism* is a name replaced in historical accounts with another name - that came later - but is more recognizable to the reader. Rohl gives the example of the Romans crossing the English Channel to invade Britain in 43 AD. There was no *English* Channel in 43 AD - because *England* did not yet exist. The term is used because practically no one knows or cares what name the Romans, or Celts, used for the channel. There are plenty of anachronisms in the Bible. Moses took an Ethiopian woman for a second wife. The name *Ethiopia* was first used by Herodotus more than a thousand years after the Exodus. In Hebrew it says Moses married a Kushite, but Ethiopia is a place name known to the reader, Kush is not. Anachronisms must simply be taken into account. They do not invalidate the Bible.

The location of Pi-Ramesse is well known. It is at Tell ed Daba in Egypt where archaeologists under the supervision of Dr. Manfred Bietak have been excavating for years. The royal city of Ramesses overlaps and lays on top of the ruins of the more ancient city of Avaris in the region known as Goshen. This was the city built and inhabited by Canaanites during the biblical time of the sojourn. It was the royal residence of the Vizier of Egypt, Joseph. The Israelites did not build the city of Ramesses because the entire Canaanite population disappeared from Avaris all at once. We will discuss the findings of Bietak at Avaris - but for now it suffices to say that once the royal city of Pi-Ramesse was built on top of Avaris, the city was forever after known by its royal name. The city of "*Raamses*" in the biblical chronicle is an anachronism. We can date Ramesses II to the post-Exodus era.

# According to the Egyptian record, Ramesses II died in his bed of old age.

# The Ramesses Exodus Theory is a mirage!

This book is devoted to putting together the pieces of the correct historical timeline, examining the archaeological and inscriptional evidence of the Exodus revealed once we look for it in the right place at the right time.

We owe Jim and Penny Caldwell for documentation of the archaeological inscriptions and other remains at Jabal al Lawz, the real Mount Sinai in Midian, Arabia. Jim Caldwell had been working for ARAMCO in Saudi Arabia for many years. Jim and Penny were avid researchers of the history and geography of the region and had visited the traditional location in the Sinai Peninsula. They left convinced it was not the correct site of Mount Sinai. In search of the true Mountain of God they followed the historical and biblical clues outlined in their book *God of the Mountain* (2008) and they were led to Jabal al Lawz in Midian.

It was David Rohl who first challenged the Orthodox Chronology. David is an archaeologist who verified there were discrepancies amounting to about two centuries in the timeline established and accepted by academicians. He discovered what the problems were with the Orthodox Timeline, and fixed them **using retro-calculation of celestial events written in the ancient records. As a result, David Rohl's New Chronology matches perfectly with the Biblical Chronology!**

# Chapter Two

## The Biblical Timeline

***For the vision is yet for an appointed time.*** *But at the end it shall speak, and not lie. Though it tarry, wait for it, because it will surely come.* (Habakkuk 2:3)

"***Chronology is the backbone of history,***" said Edwin Thiele in 1983 in his authoritative volume on the biblical timeline. Without a correct chronology, ancient history degenerates into a convoluted, confusing mess of guesstimation. The cause and effect of events is lost. Libraries are full of conflicting versions of history. **This is the state of the mainstream timeline of ancient history today! It remains a huge puzzle of a million pieces.** Many new pieces are now on the table and many more have been connected to show significant detail here and there but the big picture is still not clear. **There is hope. The New Chronology (NC) is aligned with biblical, astronomical, inscriptional, textual and archaeological evidence.**

Thanks to the New Chronology we can now ascertain the date of Joseph's arrival in Egypt. So, who was the Pharaoh who made Joseph his Vizier? That Pharaoh was Amenemhat III in whose reign accurate records were made of the lunar cycle, new moons and full moons. **Using astronomical retro-calculation we can ascertain when these lunar cycles happened, giving us an anchor, a fixed date, for the reign of Amenemhat III, the Pharaoh of Joseph.**

Having a fixed date for Amenemhat III we can identify his Vizier from the Egyptian records. They chronicle the amazing events of the Joseph story, the seven years of plenty followed by seven years of famine. The Egyptians under Amenemhat III recorded the levels of the Nile which rose for seven years providing abundance for the black land. But the Nile continued to rise to extreme flood levels preventing planting and leading to famine. It was Joseph who engineered a diversion of the flood waters into the Fayum Basin, taming the Nile, and saving Egypt from starvation. The whole of Joseph's story now comes to light.

**In our discussion of the New Chronology, we spoke of the tools one uses to make sense of the timeline of ancient history.** We have certain ***king lists***, segments of pieces that fit together. These segments, for example - a list of the genealogy of the kings of Byblos - are floating. We do not necessarily know exactly where they fit in the timeline. **We need to find an *anchor* in that sequence, a fixed point in history that can be established by *syncronizing* it conclusively to other archaeological, inscriptional or astronomical data, events or kings.**

Most of the cultures in our scenario were in a constant flux of conflict. Civilizations rose and fell. The balance of power shifted dramatically from one empire to another and then back again. Barbarian tribes swept across the boundaries of civilization in waves - came into contact, or conflict, with the civilized cultures, often becoming co-opted by them as laborers or mercenaries. When the major culture underwent a period of weakness the barbarians became their conquerors - themselves to be ejected in time by the resurgence of the major culture, conquered by others, or civilized and assimilated.

Sorting these historical dramas and putting them all together into a mosaic of ancient history is the task of historical linguists like myself. As such, I can tell you how it works from the inside perspective. **The chronological timeline of Egypt, the most extensive in ancient history, is like a jig-saw puzzle of perhaps 250,000 pieces.** Add in the pieces of the puzzle of the other Mediterranean and Mesopotamia cultures and you easily have a puzzle of a million pieces. Those puzzle pieces are scattered across a great worktable with many experts crowded around the pieces trying to put them together. **The major problem is that so many of the pieces of the grand mosaic are missing!**

> As a matter of fact, the more we pursue our studies, the less we are impressed by the dates which initially filled us with respect. We begin to recognize the framework of chronological history for what it is - a purely hypothetical structure, and one which threatens to come apart at every joint...crooked and tottering... When we reach this point in our studies, we begin to be doubtful of every single date!
>
> (C.W. Ceram, 1955: pp.133-4, *Secret of the Hittites*)

# The Critical Problem of Chronology

**One stepping stone is the process is finding *synchronisms.* For example, an inscription on a statue of the ruling pharaoh of Egypt - records that it is given to a certain King of Byblos as a gift and voila... you have a synchronism that ties that segment of the Byblian king list to the pharaohs of the Egyptian dynasty at the same time that specific pharaoh ruled.** The two segments are still floating along the timeline but now linked together. A dynasty list often seems a bit like a slide rule where you can move one scale up and down alongside other scales. If two king lists are synchronized and you can establish an anchor for one segment allowing us to date it then you can also date the second segment linked to it. Synchronizations are the key to chronology.

There lies another problem when dealing with the names of kings and royal princes, they have many names and titles. A new king will assume a new *royal* name when ascending the throne. The plethora of names and titles used in formal address are so unwieldy that shorter abbreviations, called *hypocoristicons*, were used in less formal writing. Added to this is the ever-present complication of translation of names. The Hebrew name for Moses is *Moshe* and in Egyptian *Mosis*, easy enough, but names are sometimes mangled in translation to the point they are unrecognizable. For example, the name for Pharaoh *Riama'shisha* is translated as *Raam'ses* in the Bible.

The speculative game that is played by most historians is to search the record for similarities between the ancients and personages in the Bible. The literature of biblical history is full of this kind of speculation. For example, Lennart Möller, author of *The Exodus Case*, did an elaborate analysis of 35 points of coincidence between the story of Moses and that of Senmut, the adopted son of Hatshepsut. "*Queen Hatshepsut... her closest... was Senmut (here assumed to be another name for Moses) ... This is a remarkable fit of timelines, connecting Moses with Senmut and Hatshepsut" (*Möller, 2008:121). These similarities are then regarded as a synchronism in order to try and fit the pieces of the historical puzzle together. However, over the millennia one will find many such similarities. These are guesstimations - usually wrong ones! This approach has created historical chaos.

# The New Chronology

**The New Chronology has nothing to do with such guessing games. First the timeline is established by retro-calculation from fixed anchor points that are known. The tentative new timeline is then synchronized with archaeological finds, astronomical records and inscriptional evidence. The last stage is to look for other historical similarities confirming the chronology matches the recorded events and personages of that historical period.**

Let us take Rohl's analysis of the Joseph story as a example (2015:77-85). If you remember Thiele's logic in chapter one then you know how he arrived at 1447 BC as the date of the Exodus from Egypt, and about 1400 BC as the entry into Canaan. In the *Septuagint*, the Greek translation of the Hebrew Bible, it says, "*Now the sojourning of the children of Israel and of their father* ***when they had dwelt in the land of Canaan and in Egypt was four hundred and thirty years*** " (Exodus 12:40). This would mean the sojourn in Egypt was only about half of that period, the starting point being when Abraham entered into Canaan. The King James Bible adopted the version of the Masoretic Hebrew text which left out "*in the land of Canaan*" therefore, by default, assigning all 430 years to the sojourn in Egypt.

So which is right, the *Septuagint,* translated in the 3rd century BC, or the King James, re-edited in 1611 AD - 2000 years later? The KJV took the verse from the Masoretic Hebrew text of the 10th century AD. The translators of the *Septuagint* had available a Hebrew text more than a millennium earlier than the Masoretic text. They also had Josephus account in *Antiquities of the Jews* (1st century AD), and the text of the Samaritan *Pentateuch*, first five books of the Bible, from the 3rd century BC.

All of these ancient sources agree the 430 years encompassed the period from Abraham's entry into Canaan until Moses' exodus from Egypt. The genealogies given in Bible sources also divide the 430 years roughly in half between Canaan and Egypt. This realization of the "*sojourn*" dating from Abraham's time is confirmed by Galatians 3:16-17. "*Now to Abraham and his seed were the promises made... the law, which was 430 years after, cannot disannul, that it should make the promise of none effect.*" Dividing the 430 years of the sojourn in half, we have 215 years before the Exodus in 1447 for the arrival of the Israelites in Egypt. Remember in the BC period we are counting backwards from year zero so we would add the 215 years to 1447, **giving us a date of *circa* 1662 BC for the start of the sojourn in Egypt**.

So, having a tentative date from the biblical record we need to look at any archaeological, astronomical or inscriptional evidence to synchronize the timeline. Thankfully we have the *Royal Canon of Turin*, a fragmentary papyrus which, nonetheless, preserved most of the 12th & 13th Dynasties of Egypt. The Bible says Joseph was 17 years old when taken to Egypt, spent 13 years in prison, worked under pharaoh during seven years of plenty, then in the second year of the famine brought Jacob and his family to Egypt beginning the sojourn in 1662 BC. Adding those years would make Joseph's entry into Egypt about 1683 BC in the first few years of the long (1680-1633 BC) reign of Amenemhat III according to the *Canon*.

## Can we verify this?

In the town of el-Lahun in Fayum, Egypt, there is a pyramid where lunar observations were made and records kept that have been uncovered from the 12th Dynasty period. **They recorded 11 sequential new moon observation dates in the regnal years 30-31 of Amenemhat III.** The same official who recorded these events was cited in office in the 40th year of his reign. Only Amenemhat III had a reign that long in the 12th Dynasty. **A third papyrus gives us another set of 28 lunar sightings in the reign of Amenemhat III and his father Senuseret III. We can retro-calculate these lunar events using astronomy software.**

David Lappin did the analysis of the Amenemhat lunar observations in 2001. The lunar month, new moon to new moon, is not exactly 30 days which makes it possible for us to track it. It is 29.53 days, so the appearance of the new moon varies between 29 and 30 days. 53% of lunar months have 30 days and 43% have 29 days. So if we have observation of a sequence of lunar months we can retro-calculate its accuracy via computer. If we get a 90% or better accuracy for that sequence then we have it pegged as the correct year. Sometimes overcast weather prevented a 100% correct observation record by the ancients.

**In the Lappin study, the Amenemhat lunar cycles matched to a degree of 97% accuracy, making the New Chronology dates of**

**Senesuret III, Amenemhat III and (Joseph) his Vizier as virtually certain.** Lappin's date for the ascension to the throne of Senesuret was 1698 BC, exactly as Rohl had calculated, and the ascension of Amenemhat as 1678 BC only 2 years different from Rohl's close approximation of 1680 BC. This is dramatic verification of the New Chronology! The traditional chronology places these events almost a century and a half earlier. Imagine how that prevents the possibility of finding accurate synchronisms in that time period.

## Joseph - Grand Vizier of Amenemhat III

Now we can turn to similarities in the historical record to see if they confirm the presence of the Vizier who gathered grain during seven years of plenty and saved Egypt during the seven years of famine that followed. Joseph was a Semite from Canaan. As the favored son of the patriarch Jacob, Joseph would have been taught writing in order to prepare him to take over the family's affairs upon Jacob's death. Joseph was a diviner of dreams. The hand of God was upon him, preparing Joseph for great accomplishment - but he was not yet ready.

*And when his brethren saw that their father loved him more than all his brethren, they hated him, and could not speak peaceably unto him. And Joseph dreamed a dream, and he told it to his brethren: and they hated him yet the more.*
(Genesis 37:4-5)

Joseph, full of cockiness and lacking all awareness, bid them listen to his wisdom about the dream of binding the sheaves of grain in the fields with his brothers, "*behold, your sheaves stood round about and made obeisance to my sheaf. And his brethren said, shall you indeed reign over us?*" (Gen 37:7-8). This angered his brothers to a murderous intensity. They beat Joseph, threw him down a well, then sold him to Midianite slave merchants for 20 shekels. While a slave in Egypt, Joseph ran the household of an official named Potiphar, a position Joseph probably obtained because he could read and write. Joseph soon suffered the betrayal of Potiphar's wife who tried to seduce him and, failing, accused him of taking her by force. Joseph was thrown into prison where he remained for 13 years.

By the hand of God, Joseph ended up administering the prison, probably in large part because he was literate. He made himself known as a diviner of dreams. When the butler of pharaoh was thrown into prison and dreamed dreams, Joseph said "*tell me them, I pray*" after qualifying that "*interpretations belong to God*" (Gen 40:8). At least Joseph seems to be sharing credit with God but he is still not ready.

Two full years later pharaoh had a dream of seven fat cows coming out of the Nile followed by seven lean and starving cows who ate the fat cows. None of his magicians could interpret this dream. The pharaoh was deeply distraught, "*his spirit was troubled*" (Gen 41:80). His butler remembered Joseph and he was brought before the royal court. This time Joseph says to pharaoh, "*It is not in me, God shall give pharaoh an answer*" (Gen 41:16). At that moment Joseph was ready for Yehovah to work through him.

Joseph predicted seven years of plenty followed by seven years of devastating famine that God will make come to pass very soon. "*Let pharaoh appoint officers over the land... and let them gather all the food of those good years... and that food shall be for store to... Egypt that the land perish not through the famine*" (Gen 41:34-36). Pharaoh was so impressed with his wisdom and skill that he made Joseph his Grand Vizier, second in power only to pharaoh himself.

**When Joseph was elevated to the position of Vizier of Egypt, one can be certain that literacy was prerequisite. An illiterate would not be put in a position of such power**. We know the biblical story but what transpired here was a truly miraculous relationship between Yehovah, Joseph and Pharaoh Amenemhat III. Rohl cites the Egyptian record of the period:

> So what did the Egyptian state do to try to alleviate the crisis in the time of Amenemhat III? Well, first Pharaoh's officials literally nationalized the agricultural estates of the great landowners known as the nomarchs (regional governors). These nomarchs had grown rich and powerful during the era of previous 12th Dynasty pharaohs, building splendid tombs for themselves at places like Beni Hassan and el-Bersha. Now, the building of these tombs ceased… (Rohl, 2015:97)
>
> At the same time, a 'Department of the People's Giving'… was established… set up to organize the gathering of crops by means of a conscript workforce and for the storage of grain in government silos. During the years of plenty… a grain tax was imposed… and a whole

> series of granaries were built to stockpile supplies for an anticipated downturn in Egypt's fortunes. That downturn arrived with a vengeance in Year 20 of Amenemhat III and the stored grain was then redistributed to those who could no longer produce crops on account of the castrophic floods and the resultant failure to replant their fields. (Rohl, 2015:98)

How can we be so certain about all of this detail? Well, much of it comes from the biblical story, "*There was famine all over the world. Then Joseph opened all the granaries and rationed out grain to the Egyptians, as the famine grew even worse in Egypt*" (Gen 41:56). The granaries are an archaeological fact. I have been to Sakkara (or Saqqara) where there are deep silos cut into the ground for storing grain. Does the Egyptian record confirm the Joseph story in the Bible?

During the reigns of Senuseret I, II, and III, two fortresses, Semna and Kumma, were built opposite each other in a narrow chokepoint of the Nile 700 miles upriver in Nubia. Their purpose was to guard against invasion from the south. **During the following reign, that of Amenemhat III these fortresses took on a new and critical function. "*They became a monitoring post for the annual flood*"** (Rohl, 2015:93).

## The '*Nilometer*' Records the Nile Flood

The cycle of life in Egypt revolved around the Nile. Every year heavy spring rainfall upriver caused the Nile to flood depositing a rich new layer of topsoil over the extensive and highly productive Nile River delta. In late June the Nile began to rise from its normal level peaking in late September before returning to its normal level. The flood would leave a high-water mark on the walls of the gorge at Ishetawy where the dual fortresses of Semna and Kumma were built. The Egyptians measured the Nile flood as a predictor of the harvest and a means of setting the grain tax. This simple '*Nilometer*' worked quite well in predicting the prosperous harvest, or lack of one, from the annual flood.

If the flood peak did not reach the 25 foot mark above the normal level, then the tax would be lowered in expectation of a reduced harvest from a smaller quantity of silt deposit and a restricted spread of irrigative flood waters. However, if the crest passed 32 feet then the tax would be increased expecting a more abundant harvest. Records of these measurements were discovered from the reign of Amenemhat III.

Starting in Year 3 of Amenemhat's reign the Nile flood began to rise for 17 years, during the final seven years (16701664 BC) it reached the highest average level of 40 feet above normal levels which caused enormous prosperity in Egypt during those years. Then disaster struck! For the next 12 years the Nile rose to disastrous flood levels almost twice that, a 69 foot flood mark. This left fields underwater during the plowing and planting season so there was no harvest at all! The Bible reflects this, "*For this is the second year there has been famine in the country, and there are still five years to come without plowing or harvest*" (Gen 45:6). Drought would not prevent farmers from plowing their fields, planting and irrigating them from the river. Only a flood lasting throughout the planting season would prevent the plowing and harvest. The Nilometer at the narrow defile of Ishetawy records the river peaking at this disastrous level for 12 years before beginning to decline.

Didn't Yehovah's prophecy say seven years of famine? Shouldn't 12 years of flood produce 12 years of famine? In actual fact, even after the 12 years of peak flooding, the Nile, although declining, remained at disastrous levels for 20 more years. So, what happened?

**Amenemhat's Vizier initiated a massive engineering project to dig a broad waterway from the Nile to divert water into the Fayum Basin creating Lake Moeris** - 420 miles in circumference. This reduced the annual flood to a manageable level - a miracle of insight and engineering! Through a series of levees and sluice gates the water could be controlled and released further downstream when needed. For this reason, the famine was broken, limited to only seven years, although the Nile was still flooding far above safe levels for years to come. Who was this miraculous Vizier who had the total trust of his pharaoh and his God?

## Ankhu - He Who Lives

**Upon taking up his authority as Vizier, "*Pharaoh named Joseph - Zaphenath Pa'aneah*" (Gen 41:45).** Kenneth Kitchen interprets the first half of Joseph's Egyptian name as a Hebraic transliteration of *djedu en ef*, which means "*who is called.*" The *dj* is a common transliteration of the /dz/ sound as exemplified in the name of Pharaoh Djoser or Zoser. Many Egyptian names are hard to transliterate into English because we do not have the same sounds in our language yet must seek to represent them in some way, shape or fashion.

The second part of his name is easier, pa means *the* and *aneah*, in Hebrew pronounced */aneakh/*, is the same as the Egyptian word "*ankh"* meaning '*life'* in both tongues. Taken together it says "*the one who lives*". "*What more appropriate name could Pharaoh have given to a former slave now his Asiatic Vizier than 'Joseph - he who is called "the one who lives"'*(Rohl, 2015:104). Remember that Joseph's father Jacob believed he had been killed and mourned him without ceasing.

***Ankhu* (*who lives*) was the short name, the hypocoristicon, of the most famous Vizier of Egypt.** Ankhu replaced a Vizier named Simontu during the reign of Amenemhat III. The Egyptian records refer to him as "*Ankhu, Overseer of the Fields*" and "*the granaries of Ankhu.*" The great fame of Ankhu reflected great accomplishments memorialized by the Egyptians in stone.

**Make no mistake about it, the actions taken by Amenemhat III could easily have provoked bloody revolution**. He dispossessed all his nomarchs, regional potentates, taking their lands and putting them into the hands of Joseph's trusted Semite shepherds who owed their loyalty to Joseph and Amenemhat. They would follow his dictates. The native Egyptian nomarchs, grown fat and greedy, would not. They had to go. It is no wonder Genesis 46:34 says, "*for every shepherd is an abomination unto the Egyptians.*" Amenemhat dispossessed his powerful native governors and put foreigners in charge of all his lands to save Egypt. "*And pharaoh spake unto Joseph... in the best of the land make your father and brethren to dwell, in the land of Goshen let them dwell, and... make them rulers over my cattle"* (Gen 47:6). The faith in Yehovah required of Amenemhat III was incredible!

# Bahr Yussef - The '*Waterway of Joseph'*

These two, Amenemhat III and his Vizier Joseph Zaphenath Pa'Ankhu, saved Egypt - and the Israelites - from an extended famine too awful to imagine. It was a measure of his pride in the enormous engineering feat they had accomplished that Pharaoh Amenemhat III erected two colossal statues of himself, one on each side of the great waterway where it enters into Lake Moeris. Moeris is quite possibly derived from his coronation name, Amenemhat Nu-Muaria. There he built the Hawara Pyramid for his tomb and a huge labyrinth flanking it - more striking than all of the architecture of the Greeks - according to Herodotus. **As for the grand waterway, it is still known to this day as *Bahr Yussef*, the '*Waterway of Joseph.'***

Scripture states that after Joseph, "*there arose up a new king over Egypt, which knew not Joseph*" (Ex 1:8). Despite the miraculous works of Joseph under Pharaoh Amenemhat III, there came a time when all of that was forgotten. The natives grew envious of the privileged position these Asiatic foreigners held upon Egyptian soil. The new Pharaoh, fearing the strength and numbers of the Hebrews, enslaved them in bitter bondage. In order to reduce their numbers he commanded of the Hebrew midwives, that "*if it be a son, then you shall kill him*" (Ex 1:16). The Hebrew midwives feared God and did not carry out Pharaoh's command claiming the Hebrew women had already given birth before they could even get to them. Yehovah blessed the midwives for their faithfulness (Exodus 1:20-21).

Failing to get the Hebrew midwives to cooperate, Pharaoh rallied all Egyptians to help eliminate the threat of the burgeoning Hebrew population. "*And Pharaoh charged all his people, saying, Every son that is born you shall cast into the river...*" (Ex 1:22). According to Bietak, the Canaanite graves at Avaris show a tremendous increase in burials of children during the later period of the city. Bietak states 65% of all the burials in that period were children under 18 months, three times the average of infant deaths in ancient societies (Rohl, 1995:271). If the Egyptian oppression of the Israelites recorded in the Bible is to be believed, we need to examine the claim of some researchers that the Egyptians did not practice such sacrifices. As we shall see, sacrifice and superstition loom large in the story of the Exodus.

# Chapter Three

## The Chronological Controversy

*The Heavens declare the glory of God,*
*And the firmament show his handiwork.*
***Day after day they utter speech,***
***And night after night display knowledge.***
***There is no language where their voice is not heard.***
(Psalms 19:2)

The miracles of the Exodus cannot be explained by nature or science. It is certainly within God's scope to use natural phenomena to provide miracles, however the parting of the Red Sea was not a tidal eddy nor were the other miracles likely to be caused by natural reasons - because from Moses first encounter with Yehovah - miracles accompanied his every step.

It is impossible to separate the truth of the writing of God from the validity of Moses and the Exodus epic. The Exodus story is the story of Moses, the messenger of God, whom Yehovah sent up against the all-powerful Pharaoh of Egypt. There are serious problems in unearthing the truth of the Exodus story.

**The chronology of ancient history of the Exodus period has never been established with any certainty. The debate rages over who were the pharaohs of the Exodus.** There were three; the pharaoh whose daughter took Moses into the Egyptian court to be raised, the pharaoh who sought Moses' life, and the pharaoh Moses confronted 40 years later when he returned to free the Israelites from bondage. None are specified in the Bible.

Some of the most significant, and miraculous, of the Exodus events are documented in non-biblical accounts; *The Admonitions of Ipuwer*, Antiquities *of the Jews* by Josephus, Artapanus' history entitled

*Concerning the Jews*, Manetho's chronicle of Egyptian history, and the El-Arish inscription. **Using the New Chronology we can identify the pharaoh and the princess who took Moses from the Nile, the pharaoh from whom Moses fled into Midian, and the pharaoh who defied Yehovah, brought the plagues upon Egypt and died at the Red Sea.**

What can archaeology tell us about this riveting story from the Bible? Most especially, when did this all come to pass? In the last chapter we discussed the work of Manfred Bietak at Avaris, the city of Joseph and Pharaoh Amenemhat III. It was Bietak who reported that in the later periods of occupation "*sixtyfive per cent of all the burials were those of children under the age of eighteen months.*" In the final phase of this occupation period in Avaris, "*analysis of the site archaeology suggests that a large part of the remaining population of the town abandoned their homes and departed from Avaris en masse*" (Rohl, 1995:279 citing Bietak, 1979). The archaeological evidence shows that at the time of the Exodus we have a Canaanite population centered in Avaris, Egypt - that ups and leaves all at once.

**So who was the daughter of pharaoh who took a child from the river to raise?** Pharaoh had ordered all new-born Hebrew male infants to be slain in an attempt to control their expanding population. To escape this end, Moses' mother put him in a basket in a Nile basin for the daughter of pharaoh to discover, as God had revealed to her. The princess took compassion on Moses, adopted him and raised him in the royal court.

There, Moses learned all the arts of the Egyptians including writing in hieroglyphs and cuneiform as Stephen states in Acts 7:22, "*Moses was learned in all the wisdom of the Egyptians, and was mighty in words and in deeds.*" One day Moses spied an Egyptian abusing a Hebrew laborer. Moses struck the Egyptian and killed him. Afterward, when pharaoh sought to kill him, "***Moses fled from the face of Pharaoh, and dwelt in the land of Midian***..." (Ex 2:15).

We have the testimony of Artapanus who wrote, in the first century AD, a history of the Jews in Egypt. After Alexander the Great conquered the known world, his general Ptolemy seized Egypt after his death along with the priceless jewel of Alexandria where Ptolemy established his dynastic court. There, Ptolemy and his descendants built the most renown library of the ancient world. Artapanus was likely given a royal commission to write the history in Greek. Although the work itself was lost along with the Alexandrian Library some of its contents survived via citation from authors like Clement of Alexandria and Eusebius.

## Ancient Records of the Exodus

**Artapanus tells us the pharaoh during Moses infancy was Palmanothes, his daughter Princess Merris adopted Moses after pulling him from the basket in the Nile.** Merris is probably a hypocoristicon, a shortened form of the longer formal royal name. Josephus, in *Antiquities of the Jews*, speaks of the same story using the name of Princess Thermutis, Merris is probably the Greek version of her name. For that matter, the name given to the child, Moses or *Mosis* in Egyptian, is also a hypocoristicon. This is a common ending on the names of Egyptian pharaohs, Thut-mosis, for example, meaning "*favored of the god Thoth.*" It seems appropriate for Moses because the syllable /mo/ is the Egyptian word for *water*, implying he was *taken from the water*. Like most royalty, his formal name was a paragraph long.

Artapanus gives us an extremely important starting point on the timeline of Moses' childhood. "*What we do learn from Artapanus, however, is that, at this time, the country was divided into smaller kingdoms ruled by different local lines of kings. This puts us in one of the so-called Intermediate Periods. The Third Intermediate Period is far too late and the First Intermediate Period too early.* ***So we may, with reasonable confidence, assume that we are dealing with the Second Intermediate Period*** *- that is the era from the mid 13th dynasty to the start of the 18th dynasty*" (Rohl, 2015:131).

During the intermediate periods, Egypt fell into chaos where the authority of the central government fell apart. The resulting power situation was confusing. The kingdom was so enormous it was divided into two main divisions, Upper Egypt (south) and Lower Egypt (north). Since the Nile flows from the south to the north, Lower Egypt was down river in the Nile delta whereas Upper Egypt was up river to the south.

A vigorous and capable pharaoh could assume the crown of both Upper and Lower Egypt. When Egypt was lacking a strong enough central government, each was governed by a different pharaoh, a co-ruler. This division of Upper and Lower Egypt led to a dynamic of power that was constantly in play. The ebb and flow of power between

Upper and Lower Egypt is an underlying factor in all their affairs and a recurrent theme in Egyptian literature.

When a potentate in another part of Egypt amassed so much power, co-rulership was a simple acknowledgement of reality. There are periods in Egyptian history when there are more than one dynasty at the same time. This regionalism often occurred during the Intermediate Periods when the formalities of power - official titles, crowns, clearly delineated boundaries - became sketchy, duplicative and confused.

## Pharaohs of the Exodus

The names of the pharaohs given by Artapanus were often distorted by their translation into Greek; "*the pharaoh who ruled south of Memphis, at the time Palmanothes ruled in the delta, was called Khenophres. And* ***it was Khenophres who married Palmanothes' daughter Merris and therefore became Moses' stepfather.*** *The name Khenophres we can identify... In fact, only one king in the entire history of Egypt bore it as his coronation name... Khaneferre Sobekhotep IV"* (Rohl, 2015:132).

| | |
|---|---|
| ***Greek:*** | ***Khe - noph - res*** |
| ***Egyptian:*** | ***Kha - nefer - re*** |

So, according to Artapanus, during the reign of Pharaoh Khaneferre Sobekhotep IV, "*Moses matured to become a popular royal prince in Egypt and commander of the army"* (Rohl, 2015:132). **Take note that the royal marriage between the pharaoh of Upper Egypt, Khaneferre Sobekhotep IV and Princess Thermutis (Merris), daughter of the pharaoh of Lower Egypt, would have re-united the country, making Sobekhotep IV the most powerful pharaoh of the 13th dynasty.** As proof of this we find colossal statues of Sobekhotep at Karnak in Upper Egypt and at Tanis in the eastern delta in Lower Egypt.

The Hebrew-Egyptian family of Joseph/Ankhu, who married Asenath - daughter of the high priest in Heliopolis - yielded great influence and affluence for Semites in Egypt. Joseph's daughter Senebhenas, "*married the brother of Queen Aya - the wife of the first king of the new [13th] dynasty - Sobekhotep I*" (Rohl, 2015:133), **who**

**probably gained the throne due the support of the great Vizier Joseph/Ankhu**. According to Rohl, Joseph's sons Manasseh, Egyptian name Resseneb, and Ephraim, Egyptian name Iymeru, also served as Viziers of Egypt. During this time - viziers were more powerful than pharaohs - thanks to Joseph's achievements.

The pharaoh of Joseph, Amenemhat III died, leaving his daughter on the throne, Sobekneferu, who reigned for four years and died without an heir, leaving Joseph/Ankhu - the miracle worker who had saved Egypt and tamed the Nile - firmly in control of the apparatus of government. In the power struggle that ensued at the end of the 12th dynasty and the initiation of the 13th, it was Joseph who held all the cards.

Later, Joseph's well-trained sons served in his post - as viziers who controlled the "*election*" and dismissal of some 22 pharaohs who each ruled briefly over the next 80 years. It was at that point a new branch of the family line of pharaohs from Thebes in Luxor, Upper Egypt, 700 kilometers up the Nile, took over. The first of this line - Sobekhotep III - was a pharaoh who literally "*did not know Joseph.*" **To the native Egyptians they were pushing back against a virtual takeover of Egypt by Semites via the viziership of Joseph (Ankhu), his two sons Manasseh (Resseneb) and Ephraim (Iymeru), and grandson Iymeru II, son of Ephraim.**

The Brooklyn *Slave Papyrus* from around the time of Sobekhotep III lists the names of slaves in an Upper Egypt estate, half of which had Semitic names. Many Semites would have taken Egyptian names by this time which may or may not have kept them from enslavement. The situation is eerily similar to that of the Messianic Jews of Spain at the time of the Inquisition and the Ashkenazic Jews of Germany at the time of the Holocaust. The Jews had grown strong and prosperous in position, power and wealth. Class envy was used as a diabolical weapon to destroy them. Clearly, by the time of Moses' birth the Hebrews had been dispossessed and enslaved. Infants died, or were slain, and buried at three times the normal death rate according to Bietak's excavation of graves in Avaris, the center of Hebrew habitation and influence on the Nile. It was a dark time for the chosen people.

Sobekhotep III was succeeded by his brother Neferhotep I. In light of the New Chronology of David Rohl, we can now identify the name and date of the pharaoh of the Exodus. Let me explain how the pieces of this puzzle come together. **The Babylonian astronomer priests were expert at reading the events of the night sky.** This way of thinking is reflected in the Bible, "*And God said, let there be lights in the firmament of the heaven to divide the day from the night;* ***and let them be for signs,*** *and for Moedim (appointed times), and for days, and years*" (Gen 1:14).

The priests recorded these celestial events and interpreted them. Over millennia there were many solar eclipses. In this case **the records speak of a double eclipse, a lunar eclipse plus a solar eclipse in the same month!** This is a double event that can be found, retro-calculated using astronomical software! The record of **the sky watchers cited a <u>lunar eclipse happening at sunrise</u>:** "*in the last watch [the eclipse] begins, and he [the moon] is seen with the Sun. His horns bend [away from] the sky*"' The crescent of the moon bent downward toward the earth. **Fourteen days later a solar eclipse is recorded:** "*On the 28th [day of the month] an eclipse is close by... observe the god [the sun] who in his eclipse was visible and disappeared...*" This **describes a <u>solar eclipse happening at sunset</u>.** Astronomical software not only confirmed this but also reconstructed videos of both eclipses on the horizon - the lunar eclipse at sunrise - the crescent of the partially eclipsed moon pointing down, and a solar eclipse 14 days later at sunset. **For the ancient Babylonians this was dramatic stuff! Their prediction: "*The destruction of Babylon is near!*"** (Rohl, 2015:73).

Lest one be tempted to chuckle at such soothsaying - in this case - it did indeed presage the fall of the Old Babylonian dynasty! "***At that point the Hittite king Murshili I... invaded Mesopotamia and sacked the city of Babylon,*** *bringing an end to Hammurabi's golden dynasty*" (Rohl, 2015:72). We have the king list for this dynasty - Samsuditana was the last king to reign and Hammurabi near the middle. The solar eclipse described above happened **on February 25, 1362 BC** fixed by astronomical retro-calculation. It is amazing to have proof this certain in the 2nd millennium BC! We have an anchor - a fixed time - for the destruction of Babylon in 1362 BC, a synchronized link between the Babylonians and Hittites at that date, and king lists for both! But wait, it gets better. The king list of Assyrian monarchs found in the library of Ashurbanipal allows us to count back from the reign of Ashurbanipal to Hammurabi's reign, 1560-1518 BC.

A king list gives the succession of kings during a dynastic era and the number of years each reigned but it does not tell us **when** this happened. A king list gives us a segment of history floating along the timeline. That is the reason the astronomical records are so crucial for anchoring these floating segments to a precise date in the chronology. Thanks to many decades of digging by archaeologists we now know the ***"where"*** of these ancient cities and cultures. Thanks to an equal amount of work by historical linguists we now know ***"what"*** their ancient inscriptions tell us. It is the ***"when"*** that has escaped us until now! That is why the **New Chronology is so important, it has been consistently correct in ascertaining the ancient timeline, including verifying the biblical dates of the Exodus and the Conquest.**

**The double eclipse of 1362 BC anchors the date of the fall of Babylon to the Hittite King Murshili I** which ended the reign of Samsuditana. This allows us to work our way back through the Babylonian King List from Samsuditana's demise to Hammurabi which gives us the date of his ascension to the throne in 1560 BC and a 42 year reign until 1518 BC. The '*orthodox*' chronology gives dates for Hammurabi roughly in the 18th century BC, that is between 1700-1800 BC, in error by 200 years.

In the 34th year of his reign, Hammurabi invaded the ancient and well-known city state of Mari on the Euphrates in what is now Syria. King Zimrilim of Mari was killed and his palace destroyed. According to the New Chronology, this synchronized event happened *circa* 1526 BC. Excavations at Mari in 1933 uncovered an archive of 25,000 cuneiform tablets. Among these tablets was a list of gifts to the great king by lesser potentates of the region. One of these gifts was a golden cup sent by King Yantin of Byblos in Canaan. We can now synchronize Zimrilim's 14 year reign because it ended at the destruction of Mari *circa* 1526 BC, and the reign of King Yantin of Byblos sometime within Zimrilim's reign of 1540-1526 BC.

So back to Egypt now... King Yantin's city-state of Byblos was a vassal state of Egypt whose pharaoh during his reign was Neferhotep I whom we have been searching for - lo these many pages!

**Neferhotep I was the brother and heir to Sobekhotep III, the pharaoh who marshaled the native Egyptians to "*overthrow*" and enslave the Hebrews, beginning their time of oppression!** The excavation of Byblos during the 1920s revealed a monument marked with the royal cartouche of Nerferhotep I given to Yantin, his vassal, which read:

> [An offering which the king of Egypt gives to] Re-Harakhte, that he may cause Horus and Re to be adored each day. [To] **The ruler of Byblos, Yantin**, living again, son of the ruler [Yakin], true of voice. (Rohl, 2015:76)

If we take a mid-point of Zimrilim's reign then we have Yantin of Byblos and Neferhotep I of Egypt both on their thrones approximately 1535 BC. Any historian who now wishes to change the dates of these kings from Hammurabi to Neferhotep I, needs to explain away all of this synchronized contact.

Fragment of the Royal Canon of Turin which lists the 13th dynasty of Egyptian pharaohs.

At this point we return to the Royal Canon of Turin, mentioned previously, which lists the pharaohs of the 13th dynasty of Egypt. For all the details of this analysis see Rohl (2015:149). Moses fled from Khenophres - Khaneferre Sobekhotep IV, his stepfather, to Midian where he was in exile for forty years returning to free the Israelites during the reign of Pharaoh Djedneferre (or Djed-hetepre) **Dudimose, spelled Tutimaos in Greek. To summarize, we have a fixed anchor date with the double eclipse of 1362 BC at the sack of Babylon, and synchronized contacts between the dynasties of Babylon, Hatti, Egypt and Byblos (and Mari) which allow us to link them into the New Chronology timeline.** Good, most all of them are players in the great saga of the Exodus.

# The Egyptians Were Smote by God

Manetho was the Egyptian historian who wrote the *Aegyptiaca* (*History of Egypt*) in Greek during the third century BC. This was under the reign of Ptolemy I and II, who were great patrons of literature, founding and filling the Alexandrian Library. Manetho relates this about the pharaoh of the Exodus:

> Tutimaos, in his reign, for what cause I know not, **God smote** us **[Egyptians]**. And, unexpectedly, from the regions of the east, invaders of obscure race marched in confidence of victory against our land. By main force they easily seized it without striking a blow; and, having overpowered the rulers of the land, they then burned our cities ruthlessly, razed to the ground the temples of the gods, and treated all the natives with a cruel hostility, massacring some and leading into slavery the wives and children of others. (Josephus citing Manetho)

**So the Egyptians were "*smote by God*" so hard that foreign invaders took the whole of Egypt without a blow!** This is impossible to envision in mighty Egypt without taking into account the havoc wreaked by the plagues. We will provide testimony of that havoc. First, these foreigner invaders were known to be the Hyksos, a name derived from *Hyk Shresu*, "*Bedouin Chiefs,*" or alternately derived as *Hekau Shoshu*, "*Shepherd Kings.*" They were Semites who had poured into the Mediterranean from the northeast. Armed with horses and chariots they made allies of the Hurrians, made war upon the Hittites, then found the Egyptians to be easy prey in the devastated aftermath of the plagues and the loss of their pharaoh at the Red Sea.

The Hyksos dynasty established in the delta was concurrent with the Theban dynasty of native Egyptians still in power in Thebes in Upper Egypt. One of the Theban pharaohs, Ahmose would rally the Egyptians to oust the Hyksos more than a century later to create the 18th dynasty. One of the longest reigning of the Shepherd Kings was Maibre Sheshi, the fifth Hyksos pharaoh who ruled in the last part of the 15th century (just prior to 1400 BC) as the Israelites were entering Canaan and tumbling the walls of Jericho. **A scarab amulet with the royal cartouche of Pharaoh Sheshi was found in the excavation of Jericho helping us to date the fall of Jericho to 1400 BC (Rohl, 1995 :p.309).**

# Ancient Historians of the Exodus

Moses, according to Artapanus, has grown into a capable military commander of the Egyptian army. In Moses' youth the Kushites from the south invaded and penetrated all the way to Memphis in the Nile Delta only a few precious leagues from the sea, splitting and overwhelming almost all of Egypt. Vivian Davies found a twenty-line inscription at Luxor depicting such an invasion in the Second Intermediate Period:

> Kush was a superpower which had the capacity to invade Egypt. It was a huge invasion, one that stirred up the entire region - a momentous event that is previously undocumented. They swept over the mountains, over the Nile, without limit… the supreme power of the Nile Valley, clearly Kush was at that time. Had they stayed to occupy Egypt, the Kushites might have eliminated it. That's how close Egypt came to extinction. (Davies, 2003 cited by Rohl, 2015:140)

It is from Josephus that we get the full version of Moses' desperate battle against the Kush. After the Kush had conquered all the way to the capitol of Lower Egypt at Memphis, Pharaoh [Khaneferre Sobekhotep IV] called for Moses, his anointed heir, to lead the armies against the Kushites. Moses joined battle at once and caught them by surprise. He defeated the Kush and chased them far up the river all the way to Meroë, past the fifth cataract of the Nile, a campaign that took 10 years to complete. This would have made Moses the military ruler of Upper Egypt, in effect, co-ruler of the country.

## Historians almost universally ignore the historical testimony of Artapanus, Josephus and Manetho in their rush to discredit the Exodus.

These written histories relate the flight of Moses to Midian, his return and the Exodus. Manetho and Artapanus both had access to the Alexandrian Library the most complete source of historical manuscripts in the ancient world.

Josephus was the primary Jewish historian of the first century, hated by his fellow countrymen because he had been a leader in the revolt against Rome but then went over to the other side and became a favorite of Titus Caesar. When Caesar took Jerusalem, he offered Josephus his choice of the spoils. Josephus tells us that, *"I also had the holy books by his concession."* Scholars interpret this to mean Josephus was given the scrolls of the Second Temple at Jerusalem as well as those from the library of Nehemiah in Herod's temple. These scrolls too are now lost. Josephus was the last witness to them.

Josephus was contemporary with Christ and made mention of Jesus' ministry. He became a Romanized Jew and a citizen of Rome. Since he had the most authoritative documents of the Jews, most of which no longer exist, **Josephus' account of their history in *Antiquities of the Jews* is considered to be the best available. Was Moses co-ruler of Egypt? Josephus says Moses led the armies of Egypt, re-conquered Upper Egypt from the Kush, and built mighty cities and temples. These were things pharaohs did.**

**Moses' fall from power was likely the result of his Hebrew blood.** He went out to witness the way Hebrew slaves were treated, seemingly an act of guilt and contrition for passing as Egyptian (Ex 2:11). When he saw an Egyptian beating a Hebrew he checked to see if anyone was looking, then struck the Egyptian, killed him, and hid his body in the sand. The next day, Moses again went among the Hebrews to break up a fight and one of them rebuked him:

> And he said, who made you a prince, and a judge over us?
> Do you intend to kill me, as you killed the Egyptian?
> And Moses feared, and said, surely this *thing* is known.
> Now when Pharaoh heard this *thing*, he sought to slay Moses.
> But Moses fled from the face of Pharaoh, and dwelt in the land of Midian. (Exodus 2:14-15)

Imagine the effrontery of a slave speaking to a powerful Egyptian in this manner! The slaves' challenge sounds like a veiled threat. It certainly had that effect on Moses. Obviously, Moses had something to hide. The question is: what was this "*thing*" that was known? It would be natural for the modern reader to imagine Pharaoh was reacting to the murder. However, we must reconsider. This was not a modern state with civil rights and sensibilities governing the sacredness of life but a cruel ancient kingdom where the ruling family held absolute power. They took the life of any one they chose.

There was another reason Pharaoh Sobekhotep IV might have sought the life of Moses. After Moses' defeat of the Kush, he was military ruler of Upper Egypt. Pharaoh must have seen the writing on the wall. From his perspective, Moses might take power any time he wanted. Pharaoh's hold on power was dependent upon Moses now, the conquering military hero who had saved his kingdom.

Sobekhotep IV had three sons of his own. The discovery that Moses was Hebrew might have changed any plans for his succession. It seems unlikely Pharaoh would accept putting a Hebrew stepson on the throne in an era of retaking the country from the Semites for native Egyptians. We are presuming Queen Thermutis (Merris) keep this knowledge from her husband for exactly that reason. One might further presume that it gave her leverage over Moses because she knew his deepest secret.

**It would not be paranoid for Pharaoh Sobekhotep IV to fear Moses - the great general who had defeated the Kushites and saved Egypt.** Removing Moses from power, as Sobekhotep probably felt he must, would be dangerous. In this context, Moses' killing of the Egyptian overseer could have served as a useful pretext. Moses had a price on his head because of his Hebrew blood.

Moses fled to Midian where he lived for forty years tending the flocks of Jethro the priest of Midian - who was also a believer in the God of Abraham. It was Jethro's daughter, Zipporah, whom Moses married. One day, a burning bush appeared to Moses upon Mount Horeb in Midian (Ex 3:1). Miraculously, the bush was not consumed by the fire and Moses went up the mountain to see this wonder. The Lord spoke to Moses from the midst of the burning bush.

God commanded Moses to return to Egypt, free the children of Israel from the yoke of slavery, and bring them back to Midian so they could worship God there on that mountain (Ex 3:12). Moses protested, saying he could not speak well. The Lord anointed his brother Aaron to be his spokesman (Ex 3:13) and **Moses agreed to go to Egypt, free the Israelites, and bring them back to worship God on Mount Horeb in the desert. wilderness of Sinai in Midian in Arabia (Ex 19:1-2).**

**Upon Moses' return the Pharaoh of Egypt was Dudimose or Dedumesiu (in Egyptian), or Tutimaos** (in Greek) as we examined in previous pages on the Royal Canon of Turin giving the 13th dynasty pharaohs. There are other references to the 13rh dynasty's last pharaoh - pharaoh of the Exodus!

> The Thirteenth Dynasty had by then reached its thirty third or thirty-fourth king, Dedumesiu I. If this king is to be identified with Manetho's TUTIMAIUS, **then it would have been during his reign that the Hyksos became rulers of Egypt.** This identification would appear to be confirmed by the fact that Dedumesiu is the last known king of the Thirteenth Dynasty in the inscriptions on Theban monuments at Thebes, Deir el-Bahri and Gebelein. (Nicolas Grimal, *History of Ancient Egypt*, 1992)

After the grim devastation of the plagues, Egypt was defenseless against the Hyksos invaders. Dudimose's son and heir was dead in the final plague and Dudimose himself was killed at the Red Sea ending the 13th dynasty. "*Dudimose or Tutimaos is the Pharaoh of the Exodus. Immanuel Velikovsky relates* ***Dudimose to Tutimose or Thom on the Stele of Pi-HaKhiroth*** " (Lyle, 2012:266). Velikovsky quotes the El-Arish inscription that Pharaoh Thom (Tutimaos) died at Pi-Khirothi in the "*whirlpool of the sea.*" Despite criticism of Velikovsky's interpretation of the stele, in this particular point he seems to be right. Pharaoh Thom died in the Red Sea at *Pi-Khirothi*, which is the linguistic equivalent of the name *Pi-Ha-Khiroth* in the Bible.

**Translation of the El-Arish inscription**

***The land was in great affliction. Evil fell on this earth... It was a great upheaval in the residence... nobody left the palace during nine days and during these nine days of upheaval there was such a tempest that neither men or the gods could see the faces of their next...***

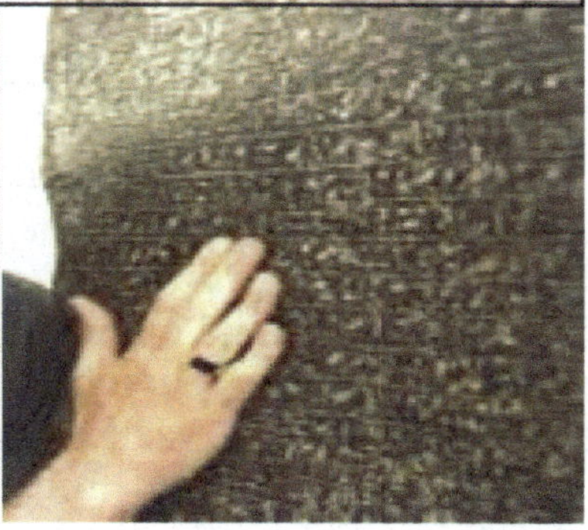

# The Pharaoh at the Red Sea

The El-Arish Shrine nicely describes the second to last plague, darkness for days. ***Pi-Ha-Khiroth*** **is named in Exodus (14:2) as the site where the parting of the Red Sea took place and the death of Pharaoh. The dedicatory monument there at El-Arish is inscribed with the royal cartouche of King Thom** which Velikovsky claims is the hypocoristicon of Tutimaos. This is supported by **Exodus 1:11** where the enslaved Israelites are building **the city of *Pi-Thom***, the "*estate of Thom.*"

The Lord had prepared Moses with weapons for the spiritual battle that was to come. The first of these was the ability to transform his staff into a serpent. The second was the appearance of a leprous hand when Moses removed his hand from his cloak. The third was the turning of water into blood as it was poured onto the land. In an extraordinary battle before Pharaoh Dudimose, Moses and Aaron performed these miracles as they confronted the priests of Egypt. Since Aaron had been appointed by God to act as Moses' spokesman, it was he who turned the rod into a serpent before Pharaoh.

*Now the magicians of Egypt, they also did in like manner...*
*For they cast down every man his rod,*
*and they became serpents:*
*but Aaron's rod swallowed up their rods.*
(Exodus 7:10-12)

Dudimose/Tutimaos still refused to liberate the Israelites. Yehovah intended the confrontation to be a complete defeat of Pharaoh, his priest magicians, and the gods of Egypt. He was not going to allow this conflict to be easy for Pharaoh Dudimose. "*I will harden Pharaoh's heart, and multiply my signs and wonders in the land of Egypt*" (Exodus 7:3). A dramatic and miraculous conclusion was destined for the whole of Egypt including Pharaoh and his army. God brought down a series of plagues upon Egypt, each worse than the last. The river Nile was turned into blood - a plague of frogs covered Egypt, followed by plagues of: lice - flies - sick cattle - boils and abscesses on man and beast - violent hailstorms - locusts - darkness for three days - and finally death of the firstborn of every Egyptian family.

# Admonitions of Apuwer

We have documentation of the plagues in an ancient papyrus of the era called the *Admonitions of Ipuwer*. This is not a work of writing decreed by pharaoh but one ascribed to the writer himself, Ipuwer. **The *Admonitions of Ipuwer* describe an Egypt in complete chaos. "*The river is blood.*"** "*Many dead are buried in the river.*" Note that it did not say the river runs with blood as one might expect from so many bodies being unceremoniously dumped into the Nile. Ipuwer said, "*the river is blood.*" "*There's blood everywhere, no shortage of death.*" Although there was no shortage of death, almost everything else was in short supply. "*Exhausted are materials [for] every kind of craft.*" Ipuwer included grain along with all kinds of wood. "*Trees are felled, branches stripped... Food is lacking... Great hunger and suffering*" (Quotes from the *Admonitions of Ipuwer*).

"*Even noblewomen go hungry.*" Having lost their husbands and their support, a commoner who was wifeless could now take a noblewoman since there were fewer males of high station left alive. "*He who slept wifeless found a noblewoman.*" "*The land is deprived of kingship. See, all the ranks, they are not in their place, like a herd that roams without a herdsman... Foreigners have become people everywhere.*" This last quote referred to the uprising of the Hebrew slaves who demanded the jewels of the wealthy. "*Poor men have become men of wealth.*" "*Gold, lapis lazuli, silver and turquoise, carnelian, amethyst... are all strung on the necks of female slaves. Gold is lacking.*"

In the end, even the most prominent magicians of Egypt were forced to admit: "***This is the finger of God***" (Ex 8:19). Pharaoh relented at last and let the Israelites leave, but Yehovah had promised Moses the Israelites would not leave empty handed. The Egyptians, terrified by God's power, gave the Hebrew slaves whatever they asked so they would leave.

*And the children of Israel did according to the word of Moses; And they borrowed of the Egyptians jewels of silver, and jewels of gold... And they spoiled the Egyptians.* (Exodus 12:35-36)

# Plagues of Egypt

This series of plagues would have devastated Egypt. Most of the men in authority died in the final plague which killed the firstborn. Typically, the firstborn took leadership posts, even scribes were chosen primarily from among the firstborn. The military were left leaderless and dysfunctional - and lawlessness resulted. The Nile River was turned to blood. The country was full of death, men and beasts died from the plagues. Harvests were destroyed by the violent hailstorm which flattened all of the crops and even the trees. What the hailstorm did not destroy was eaten by the plague of locusts.

The plagues of frogs, flies, and lice spread disease everywhere. So, in the end, the plagues became one of physical disease as well, causing the Egyptians to give in to the supremacy of Yehovah. They not only allowed their slaves to leave, but also gave them gold, silver, precious cloth - whatever they wanted - fearing the wrath of God.

Within days Pharaoh Dudimose's ego had recovered sufficiently to order his army to pursue the fleeing Israelites. The administrative apparatus of Egypt went with Pharaoh in pursuit of the Hebrews. It was perilous to hand over authority to a regent out of fear a king might return to find his kingdom usurped. Therefore, after the Red Sea finale not only was Pharaoh dead but also the most trained and trusted core of civil servants who comprised his mobile government. **The plagues plus the final destruction of the army at the Red Sea erased much of the upper class in Egypt, including Pharaoh's heir, his first and only son. Egypt was leaderless!**

Ipuwer reports "*We don't know what has happened in the land.*" At the time, the Egyptians knew nothing of what had happened at the Red Sea. The Bible said there was not a single survivor. The Egyptians were waiting in limbo for a leader and an army that never returned. This lack of knowledge of what happened prevented them from seeking new leaders and taking steps that would have ended the chaos. There was practically no one left with any authority. **Not only were their leaders fallen but their gods as well. "*What shall we do?*" "*If I knew where God is I would serve him.*"** (quotations - *Admonitions of Ipuwer*).

# Chapter Four

## Archaeo-Astronomy

*Yea, the stork in the heaven knows her appointed time...*
*But my people know not the judgment of Yehovah!*
(Jeremiah 8:7)

**The biblical timeline of the Exodus is well-established, from numerous passages in Scripture, to within a year. The Exodus from Egypt happened in 1446 BC.** Modern historians have moved that up 200 years to the 13th century. How could this happen? Historians have taken two passages from the Bible and interpreted them as evidence the Exodus happened in the 13th century BC rather than the15th. The first passage is Exodus 1:11 when the Egyptians enslaved the Israelites of the sojourn, descendants of Jacob, to build the royal estates of "*PiThom and Raamses.*" For this reason, the pharaoh of the Exodus is identified as Ramesses the Great and therefore the Exodus happened in his reign from 1279-1213 BC - the 13th century.

Historians confirm the *Ramesses Exodus Theory* by a second verse in 2nd Chronicles (12:2-9) when Jerusalem is conquered by King Shishak of Egypt. Historians assume this must be Pharaoh Shoshenk of the Egyptian record who did indeed invade Israel. "*And it came to pass, that in the fifth year of King Rehoboam, Shishak king of Egypt came up against Jerusalem.*" Rehoboam, the son of Solomon who died in 930 BC, surrendered Jerusalem to Shishak in 925 BC. It seems to confirm the 13th century as the time of the Exodus and the Conquest of Canaan.

This is the mainstream position of modern archaeologists, historians and Bible scholars. **There are only a couple of problems with this mainstream chronology of the Exodus. Ramesses was not the pharaoh of the Exodus - and Shishak was not Shoshenk!**

*The Ramesses Exodus Theory* has been created using two verses in a way that completely contradicts the biblical chronology of the Exodus! As a result, researchers looking for evidence of the Exodus from Egypt and the conquest of Canaan in the 13th century find nothing. There is no evidence to show the Israelites were in Egypt then, nor that any city of Canaan, such as Jericho, was destroyed at that time.

One might think this would prompt historians to recheck their premises and evidence for misinterpretation. In fact, quite the opposite has been the result. Secular scholars have triumphantly declared the demise of the Bible as an accurate record of history, the Bible stories are a myth! "*We can say now with considerable confidence the Bible is not a story of anyone's past.*"

**What was actually needed was a rejection of the baseless 19th century assumption that the Pharaoh was Ramesses II. That catastrophic error and the equally damaging misidentification of Shoshenk I with the biblical Shishak are the two guilty parties here... not the biblical narrative.**

The confusion of Shishak with Shoshenk actually dates to Champollion, '*the father of Egyptology,*' who - in 1822 - cracked the code of the Egyptian hieroglyphs, which had been a dead language for many centuries. Despite this magnificent achievement, when Champollion went to Egypt in 1829, he mistranslated the record of Pharaoh Shoshenk on the walls of the Temple of Karnak where it glorified his campaign into northern Israel and the capture of her cities.

> Out from the pages of the Old Testament - Ajalon, Gibeon, Mahanaim, Rehob, Beit Shean, Shunem, Taanach, Megiddo. Then Champollion came across a name-ring that caused great excitement. He began to read the glyphs: Iuda ha Malkuth - 'Judah the Kingdom'. Here was a key link to the Bible.

To Champollion this was, beyond doubt, the Shishak of the Bible! In fact, it was not. By 1888, Wilhelm Müller had already proven Champollion's hasty translation of "*Iuda ha Malkuth*" (Judah the Kingdom) was wrong. It read "*Yad ha Melek*" (Hand of the King) a military outpost near Megiddo. **Jerusalem is not listed on Shoshenk's conquest record at Karnak!** None of the cities on his list, except for Ajalon which he passed through on the way north, were in the southern kingdom of Judah. He did not conquer Jerusalem. Shoshenk's campaign went into northern Israel and had an entirely different target. How could he have conquered Judah when he wasn't even there?

*"The kingdom of Aram-Damascus controlled the upper Jordan Valley and significant areas in north-eastern Israel and devastated major Israelite administrative centers in the fertile valley as well."*[7] Egypt, having regained its hegemony over Israel and Judah was now seeing that threatened in the north by King Hazael of Aram. Shoshenk's goal was to defend his allies in the northern kingdom of Israel, not to destroy them. **The campaign of Shoshenk was to save Israel from the Arameans who were taking them over.**

> **And the anger of Yehovah was kindled against Israel, and he delivered them into the hand of Hazael, king of Aram... and Jehoahaz besought Yehovah, and Yehovah hearkened unto him, for he saw the oppression of Israel, because the king of Aram oppressed them. And Yehovah gave Israel a savior, so that they went out from under the hand of the Arameans.**

The "savior" of Israel was Shoshenk I and his army who *"took it upon himself to throw out the plundering hordes of Aram Damascus from the Northern Kingdom of Israel... Shoshenk retook the cities which once belonged within the New Kingdom empire and, as a result... he became Israel's 'savior' from the oppression of the Arameans."* The '*burnt*' levels in the archaeological records of that era in Israel are due to the devastation of the Arameans, not the Egyptians. **This Shoshenk, the '*savior*' of Israel - was not Shishak, the '*scourge*' of Israel in the Bible!**

Israel was a minor player in the cradle of civilization. The mighty empires of the Egyptians, Hittites, Babylonians, Assyrians, Greeks and Romans surrounded Israel and often fought over her. As a critical crossroad between these trading partners and ambitious rulers, Israel was almost constantly under the hegemony of one or the other. Campaigns of conquest by the Egyptians (and others) into or through its territory were common.

One of these campaigns was led by none other than Pharaoh Ramesses II. The Egyptian record of that campaign is on the north pylon of the Ramesseum at Thebes - the conquest monument of Ramesses the Great - which states:

**The city which the king [Ramesses II] plundered in year 8 - Shalem**
[brackets mine]

*Shalem*, or alternately *Salem*, from the root word *Shalom* meaning peace, was the original name given to *Yeru-shalaim* [Jerusalem] in the Bible, the /s/ and /sh/ sounds being regional variations in the Semitic dialects of the era. In Genesis 14:18, "*Melchizedek king of Shalem... the priest of the Most High God*" blessed Abraham by the hand of Yehovah.

Ancient kings had many names and titles so it was commonplace to use a hypocoristicon, a short name. One might think of it as a nickname **but it was really a short formal name used in the official record. In Egyptian, '*Ramesses'* is actually pronounced '*Riama-Shisha*', thus the hypocoristicon *Shisha*, in Hebrew *Shishak*, was the short name of Ramesses.** Of this there is sufficient evidence since Ramesses is referred to as *Shisha, "four times in the document... Papyrus Anastasi I.*"

Although historians disagree about the date of his reign there is no dispute that Ramesses II (Shishak) made a campaign into Canaan. Kenneth Kitchen is the expert on the Ramesses campaign into Canaan during year 8 of his reign:

> Ramesses took things in hand. In no time he reached Gaza (1) ... A flying column swept the Shoshu (Bedouin) back east, out of Canaan entirely (2). Then Ramesses dealt with East Palestine... the flying column struck down the Negev hills (3), across the Rift Valley south of the Dead Sea (4), and up into Edom-Seir (5) ... swung north... into the heartland of Moab... to conquer Butartu (6) ... At the same time, Ramesses himself swept round in a clockwise arc to complete the pincer movement - across the hilly central ridge of Canaan past **Jerusalem** (7) **[which surrendered] ...** over the Jordan, past Jericho (8) ... striking at Dibon (9). [brackets mine]

I have always considered it amusing that secular historians insist Ramesses the Great was the biblical pharaoh of the Exodus when - according to the Egyptian records - he campaigned in Canaan long after the Exodus and conquest, reigned for 66 years and died of old age in his bed. **Without a doubt, Ramesses the Great is Pharaoh Shishak who is cited in 2nd Chronicles, but he is not the pharaoh of the Exodus!** Ramesses reigned in the 10th century BC not the 15th century.

So, what about Exodus 1:11 where the Israelites are building the royal estates of *Pi-Thom and Raamses*? The Israelites built Avaris, royal city of Amenemhat III and his Vizier Joseph. Later, other pharaohs built there, including Pharaohs Thom and Ramesses. Once it became the royal city of Ramesses the Great it was always and forevermore referred to by *Pi-Ramesses*.

**David Rohl pointed out the reference to the royal city of Pi-Raamses is an *anachronism* - a modern name used in historical accounts in place of an ancient name - because the modern name is more recognizable to readers.** There are plenty of anachronisms in the Bible. For example, it says Moses took an Ethiopian for a wife (Num 12:1). *Ethiopia* was first used by Herodotus a thousand years after the Exodus. In Hebrew it says Moses married a Kushite, but Ethiopia is a place name known to the reader, Kush is not. Anachronisms must simply be taken into account. They do not change the timeline of the Bible.

Modern historians have agreed upon a timeline of the ancient world - especially as it relates to the Exodus. It just does not work. There is a constant conflict between the Orthodox Chronology (OC) and the evidence of history. It does not gel.

Nonetheless, historians have used this timeline to declare the Bible is not an accurate record of history. "*Virtually every archaeologist and historian working in Egypt and the Holy Land these days is convinced that the great events described in the story of the Israelite Exodus from Egypt never happened.*" If the conventional timeline is correct all historical discoveries should fall into place and validate the accepted chronology. That is not happening! But it will take more than simply pointing out the conflicts of evidence to change things. It will take a restructuring of the chronology! David Rohl is the expert who took on the epic task of recalibrating the timeline of ancient history!

Rohl began his journey as an Egyptologist by noticing there were errors in the Egyptian chronology upon which all timelines were based. Over a lifetime of professional research, he found the problems in the Egyptian record, corrected them and provided evidence to justify establishing a new timeline, a New Chronology (NC). **David Rohl's New Chronology turned out to be a perfect fit with the biblical timeline!**

Rohl used a dramatic new technology that is revolutionizing the accuracy of ancient history, archaeoastronomy. The field of archaeoastronomy is so new it was only given its name in 1973 by Elizabeth Chesley Baity, a pioneer in the field of using astronomy to understand the archaeological evidence of such structures as Stonehenge and the Pyramids. "*Archaeoastronomy is a form of information recovery with time - and space - specific aspects which provide... a new theoretical framework for explaining certain problems of prehistory*"(p.390). "*We have seen that astronomers can now compute the objective astra facts for any culture studied*" (p.422).

**In short, using archaeoastronomy, these ancient sites set up to study celestial events of the heavens can often tell us - not only about <u>what</u> they recorded - but <u>when</u>! An eclipse, for example, can be retro-calculated by astronomical computer software giving us the exact date of the ancient event if it is described precisely enough.**

**To establish a correct chronology, one must begin with fixed *<u>anchor points</u>*, dates that can be scientifically proven to be accurate.** There are great swaths of history that are dated by the methods of the ancient world. **They are dated by *<u>king lists</u>*, almost every major culture made a record of dynastic succession of their kings and how long each reigned.** When they recorded an historical event or a celestial sighting in the annals, it would be dated to the year of the reign of the current king.

We have already seen an example of this from the Bible, "*And it came to pass, that in the fifth year of King Rehoboam, Shishak king of Egypt came up against Jerusalem" (1st Kings 14:25).* **This demonstrates the third crucial concept of chronology - *<u>synchronism</u>*.** This shared event chronologically ties the reign of Rehoboam of Judah to the reign of Shishak king of Egypt.

Therefore, Rehoboam and Shishak are contemporaries - which links the timelines of Judah and Egypt together at that moment in time. **If we know the correct date of that event we can now count backwards, and forwards, on their king lists to ascertain the absolute dates of each king of both nations.** That is, *if* we have the anchor, the correct date of the shared event! If there are no *anchors,* the *king lists* are simply segments of royal succession floating along the timeline not fixed to any date.

As a chronologist I can give you the inside view of the process. It is all based on Egypt, the longest and most extensive record. There were three periods of unified authority over Egypt - the Old Kingdom, Middle Kingdom and New Kingdom. In between these well-recorded periods of high civilization and accomplishment, there are the '*intermediate periods'* when central authority collapsed, multiple regional potentates took power, and the historical record is far less certain. So, following the Old Kingdom was the First Intermediate Period (FIP), after the Middle Kingdom was the Second Intermediate Period (SIP), and after the New Kingdom we have the Third Intermediate Period (TIP). The king lists of the well-recorded dynasties give a fairly reliable incremental passage of time and pharaohs. **These *king lists* are like yardsticks connected by pieces of elastic, the *intermediate periods,* which, because of their spotty and chaotic records, can be stretched or squeezed to fit the timeline into whatever theory historians wish to expound. Keep that in mind.**

**The sacking of Thebes in Egypt by the Assyrian King Ashurbanipal in 664 BC is the earliest historical event whose date is universally accepted by all historians.** It synchronizes the 38th year of Ashurbanipal to the last (26th) year of the reign of Pharaoh Taharka. Taharka was the last of the Nubian kings who had ruled over Upper Egypt since Usermatre Piye invaded in 732 BC. We know that by counting back on the Egyptian king list from the anchor date of 664 BC. Can we verify this date?

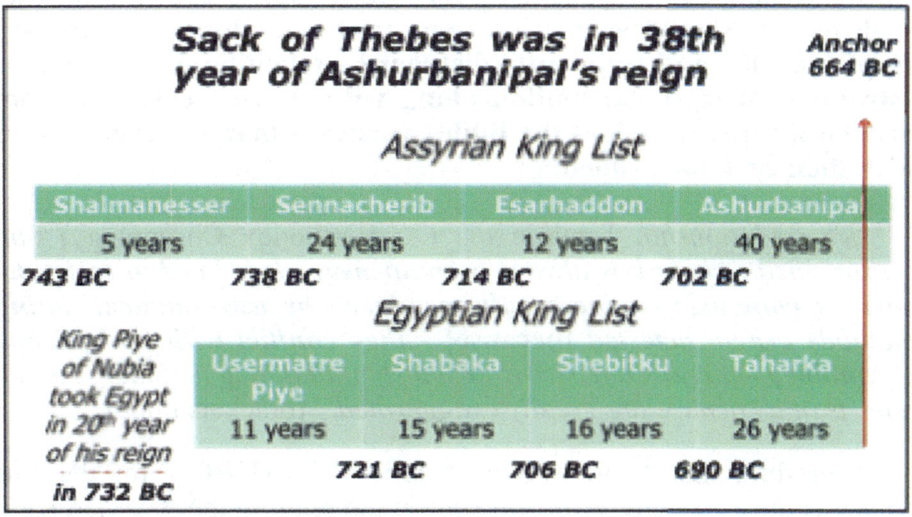


The Assyrians of Nineveh held to the long-established practice of recording the events in the heavens for prophetic signs. This was a biblical practice as well, "*And God said, let there be lights in the firmament of the heaven to divide the day from the night; and let them be for signs, and* ***for appointed times****, and for days, and years*" (Gen 1:14). These astronomical prophet-priests were called Eponyms and they searched the skies constantly. One of these, named Bur Sagale, recorded in the 9th year of Ashur-Dan III, "*Revolt in the city of Assur* (Nineveh). *In the month Simanu an eclipse of the sun took place.*" Jonah was sent by God to Nineveh, a great city sunk in iniquity. **Jonah arrived - there was an eclipse, there were riots, there were plagues and the people of Nineveh heeded the warning and they repented! Yehovah stayed the destruction of Nineveh!** (Jonah 3:10)

**Picture of the Bur-Sagale eclipse retro-calculated and imaged by the astronomy software Redshift.**

**June 15, 763 BC is, therefore, the 9th year of Assyrian King Ashur-Dan III.**

The New Chronology (NC) has been shaped by astronomical retro-calculation in conjunction with archaeological and inscriptional evidence. **It should surprise no one that orthodox historians are suspicious of - and frequently disregard - scientific dating such as astronomical retro-calculation along with written evidence from historical sources such as the Bible, whenever that evidence clashes with their orthodox timeline!**

"*The relationship between the OC* [Orthodox Chronology] *and astronomical dating has always been uneasy...* ***It is hard to resist the obvious conclusion... the results produced by astronomical dating methods are so detested*** **[because]** ***- they conflict with a cherished chronology -*** *there is a desire to 'rubbish' that dating method. But the reports of its demise are greatly exaggerated.*" [brackets mine]

Therefore, the absolute date of June 15, 763 BC - for the Bur Sagale eclipse - would be the 9th year of the reign of the Assyrian King Ashur-Dan III. That date conflicts with the Orthodox Chronology (OC). However, if we count forward on the Assyrian king list we arrive, one century later, to Ashurbanipal and the sack of Thebes in

664 BC - the earliest date accepted by all historians, and now validated by astronomical retro-calculation. **This is scientific evidence of the correct ancient timeline that cannot be dismissed or wished away!**

Both the Assyrian and the Babylonian chronicles confirm this 664 BC event. The ancient Egyptian historian Manetho confirmed the date. He lived in the 3rd century BC in Alexandria and had their great library as a resource. Also, the 3rd century BC Chaldean priest Berosus wrote, in Greek, of the conquest happening then, citing the event from Hammurabi's library in Babylon. Surviving monuments also confirm the dates of the reign of both these kings as do the analyses of archaeological evidence.

Edwin Thiele started with this astronomical retro-calculation of the Bur-Sagale eclipse on June 15, 763 BC (9th year of Ashur-Dan III). Thiele then counted back in the Assyrian king list to Shalmaneser III giving us a date for his reign from 858-824 BC. This was the king who, in the 6th year of his reign, marched into Canaan to fight a confederacy of kings, one of which was King Ahab of Israel (1st Kings), husband of Jezebel, who sent 2000 chariots and 10,000 warriors into battle against Shalmaneser at Karkar, where Ahab was killed. **King Ahab's defeat in 853 BC anchors a scientifically fixed date which is confirmed in the biblical chronology!**

Thiele could then calculate Ahab's reign from 874-853 BC, and from sources outside the Bible fix the battle of Karkar in the last year of his reign. If we count back in the biblical chronology of kings of the divided monarchy of Judah and Israel, that brings us to 970 BC as the year Solomon ascended the throne for his 40-year reign and therefore 930 BC, the year of his death, as the date of the '*schism*' when the country was divided into a southern kingdom of Judah under the rule of Solomon's son Rehoboam, and the northern kingdom of Israel where a rebellious former officer of Solomon, Jeroboam, seized power.

**The important point is that the New Chronology [NC] matches the biblical chronology!** Continuing backward in time from the dates given in 1st Kings 6:1, "*480 years after the Exodus*" Solomon began building the Temple in the 4th year of his reign. This gives us a date of 1447 BC for the Exodus from Egypt. This is confirmed by the dates given in Judges 11:26 where Jephthah (*circa* 1100 BC) states the Israelites began the conquest of Canaan 300 years previously which would give us 1400 BC. Now we have a date for the onset of the Conquest of

Canaan which followed the Exodus and the 40 years of wandering. We anchored the later date of the epoch to the Bur Sagale eclipse. Can we anchor the early end with astronomical dating?

In fact, we have already established an earlier date in the biblical chronology. In chapter one we discussed David Lappin's analysis of the 39 lunar cycles recorded during the reign of Joseph's pharaoh, Amenemhat III and his father Senuseret III, which matched the NC calculation to a degree of 97% accuracy. The ascension of Amenemhat III to the throne was calculated by Lappin as 1678 BC. **This makes the NC and biblical dates of Amenemhat III and Joseph, his Vizier, perfectly aligned and scientifically certain!** This is still almost two centuries from the Exodus. Can we verify a closer date? Indeed, we can.

The Babylonians recorded a double eclipse, a lunar eclipse at sunrise and two weeks later, a solar eclipse at sunset. A lunar and a solar eclipse in the same month - such a distinctive sighting is beyond argument! The lunar eclipse was retro-calculated to February 25, 1362 BC at sunrise and the solar eclipse 14 days later was March 12, 1362 BC at sunset. The priests prophesied the fall of Babylon, which might tempt one to chuckle except "*at that point the Hititte king Murshili I... invaded Mesopotamia and sacked the city of Babylon.*" **Whatever one may think of prophetic signs, the facts are that Murshili I invaded Babylon in 1362 BC ending the reign of King Samsuditana, giving us a fixed anchor within a few decades of the Israelites entering Canaan, and a synchronized link between the Hitittes and Babylonians.** Let's continue the process a little further and see what else it **shows.** If we are on the right path then our New Chronology and biblical timeline will sync with the evidence.

Continuing to count our way back through the Babylonian king list gives us a date of 1560-1512 BC for the reign of Hammurabi, the famous lawgiver. Hammurabi invaded the city-state of Mari, killing King Zimrilim, *circa* 1526 BC. Excavations at Mari uncovered 25,000 cuneiform tablets. One of these tablets recorded a golden cup sent by King Yantin of Byblos to King Zimrilim sometime within his reign from 1540-1526 BC. Byblos, just north of Israel, was a vassal city-state of Egypt whose pharaoh during King Yantin's reign was recorded as Neferhotep I.

Archaeological excavation of Byblos revealed a monument, marked with the royal cartouche of Neferhotep I, given to Yantin, which reads, "[To] *the ruler of Byblos, Yantin, living again, son of the ruler [Yakin], true of voice*" (Rohl, 2015:76). Taking the midpoint of our date of Zimrilim's reign - we have Yantin of Byblos and Neferhotep I of Egypt both on their thrones approximately 1535 BC, a generation before the birth of Moses.

To summarize, we have a fixed anchor date with the double eclipse of 1362 BC at the sack of Babylon, and synchronized contacts between the dynasties of Babylon, Hatti, Byblos and the 13th dynasty of Pharaoh Neferhotep I of Egypt allowing us to anchor the New Chronology timeline of the Exodus.

It is so exceedingly strange that historians claim there is no evidence, outside of the Bible, of the Exodus because the story is told in quite some detail in the writing of three very renown ancient historians: Artapanus, Manetho and Josephus. I am weaving their stories into the documentary evidence we have of the 13th dynasty. **The Royal Canon of Turin lists the pharaohs of the 13th dynasty of Egypt.**

Fragment of the Royal Canon of Turin which lists the 13th dynasty of Egypt

According to these ancient sources, Neferhotep I was the brother and heir to Sobekhotep III, the pharaoh who marshaled the native Egyptians to '*overthrow*' and enslave the Hebrews, beginning their oppression. **Neferhotep was succeeded by Khenophres - the Greek name for - Khaneferre Sobekhotep IV, the stepfather of Moses.** Moses fled from this pharaoh to Midian where he was in exile for forty years - returning to free the Israelites during the reign of Pharaoh Djed-neferre, or Dudimose, spelled Tutimaos in Greek.

Moses, according to Artapanus, had grown into a superb military commander of the Egyptian army. When Moses was a young man the Kushites from the south invaded and penetrated all the way to Memphis, splitting and overwhelming almost all of Egypt. Vivian Davies found a twenty-line inscription at Luxor depicting that invasion in the Second Intermediate Period:

Kush was a superpower which had the capacity to invade Egypt. It was a huge invasion, one that stirred up the entire region - a momentous event that is previously undocumented. They swept over the mountains, over the Nile, without limit… the supreme power of the Nile Valley, clearly Kush was at that time. Had they stayed to occupy Egypt, the Kushites might have eliminated it.

It is from Josephus that we get the full version of Moses' desperate battle against the Kush. After the Kush had conquered all the way to the capitol of Lower Egypt at Memphis, Pharaoh Khenophres [Khaneferre Sobekhotep IV] called for Moses, his anointed heir, to lead the armies against the Kushites. Moses defeated the Kush and chased them far up the river all the way to Meroë, past the fifth cataract of the Nile, a campaign of ten years.

Forty years later the Egyptians faced another crisis. Moses, after his exile, returned to lead the Israelites out of Egypt. Pharaoh Dudimose [in Greek - Tutimaos] was on the throne at that time. We have a record of those events from Manetho, the Egyptian historian who wrote the *Aegyptiaca* (History of Egypt) in Greek during the third century BC.

Manetho relates this about the pharaoh of the Exodus:

> **Tutimaos, in his reign, for what cause I know not, God smote us [Egyptians].** And, unexpectedly, from the regions of the east, invaders of obscure race marched in confidence of victory against our land. By main force they easily seized it without striking a blow; and, having overpowered the rulers of the land, they then burned our cities ruthlessly, razed to the ground the temples of the gods, and treated all the natives with a cruel hostility, massacring some and leading into slavery the wives and children of others. [brackets mine]

The Egyptians were "*smote by God*" so hard that foreign invaders were able to take the whole of Egypt without a blow? **The invaders encountered an Egypt in complete chaos from the plagues brought down by Moses - as described in the *Admonitions of Ipuwer!*** "*The river is blood.*" "*Many dead are buried in the river.*" It did not say the river runs with blood as one might expect from bodies being dumped into the Nile. Ipuwer said, "*the river is blood.*" "*There's blood everywhere, no shortage of death.*" Although there was no shortage of death, almost everything else was in short supply due to the plague of hail. "*Exhausted are materials [for] every kind of craft...*" "*Trees are felled,* *branches stripped... Food is lacking... Great hunger and suffering.*" "*Even noblewomen go hungry.*" A commoner could now take a noblewoman since there were few males of high station left alive. "*He who slept wifeless found a noblewoman.*" "*The land is deprived of kingship. See, all the ranks, they are not in their place, like a herd that roams without a herdsman... Foreigners have become people everywhere.*"[21]

Moses directed the enslaved Hebrews to rise up and demand the wealth of the Egyptians. "*Poor men have become men of wealth.*" "*Gold, lapis lazuli, silver and turquoise, carnelian, amethyst... are all strung on the necks of female slaves. Gold is lacking.*"

*And the children of Israel did according to the word of Moses;*
*And they borrowed of the Egyptians jewels of silver, and*
*jewels of gold... And they spoiled the Egyptians.*
(Exodus 12:35-36)

Foreign invaders taking mighty Egypt without a blow would have been impossible without the havoc wreaked by the plagues preceding the Exodus - recorded in the Bible and the *Admonitions of Ipuwer.* These foreign invaders were the Hyksos, a name derived from *Hyk Shresu*, "*Bedouin Chiefs,*" also known as *Hekau Shoshu*, "*Shepherd Kings.*" They poured into the Levant from the northeast. David Rohl identifies these Hyksos as the Amalekites who attacked the Israelites in force, then found the Egyptians easier prey in the chaotic aftermath of the plagues and the loss of their army and pharaoh at the Red Sea.

One of the longest reigning of these Shepherd Kings was Maibre Sheshi, the fifth Hyksos pharaoh who ruled in the last part of the 15th century (just prior to 1400 BC) when the Israelites entered Canaan and tumbled the walls of Jericho. A scarab seal with the royal cartouche of Pharaoh Sheshi was found in the excavation which dates the fall of Jericho to 1400 BC. The Hyksos dynasty established in the delta was concurrent with the Theban dynasty of native Egyptians still in power in Thebes in Upper Egypt. The Theban pharaohs, twin brothers Kamose & Ahmose - more than a century later - would rally the Egyptians to oust the Hyksos. When Kamose was killed in battle, Ahmose took the throne and finished the job.

**In their rush to discredit the Exodus - historians typically ignore the testimony of Artapanus, Josephus, Manetho and Ipuwer.** These

**Seal of Pharaoh Sheshi, 1400 BC, found at Jericho.**

authors' written histories relate the flight of Moses to Midian, his return and the Exodus. Manetho and Artapanus both had access to the Alexandrian Library, the most complete source of historical manuscripts in the ancient world. Josephus was the primary Jewish historian of the first century, and a favorite of Titus Caesar. When Caesar took the city of Jerusalem, he offered Josephus his choice of the spoils. Josephus wrote, "*I also had the holy books by his concession.*" These ancient Hebrew histories are now lost, as is the Alexandrian Library. **These men were the last witnesses to these archives. Their sources were far better than ours are today!**

# Chapter Five

## Grimme's Tales

***How much better is it to get wisdom than gold!***
**(Proverbs 16:16)**

The Queen of Sheba traveled to Israel in the 10th century BC to seek out the wisdom of King Solomon, son of King David. The land of Sheba was at its cultural and commercial apex, a mighty trading nation and center of culture and literacy. The story of Solomon and Sheba was recorded in both the Bible and the *Kebra Negast*. The *Kebra Negast* is the national epic of the Ethiopians who are the modern-day descendents of the ancient Sabean kingdom. According to Sabean legend, Sheba and Solomon had a son together, Manelik, who traveled to Israel upon achieving manhood and took the Ark of the Covenant back with him when he returned to Sheba.

Most historians consider the story of Manelik to be fiction. Many also consider Solomon and Sheba themselves to be myth, citing the fact that there is no certain physical evidence of either of them outside of the written record. Readers can look at the historical record of King Solomon and the Queen of Sheba and judge for themselves. **The results of archaeo-astronomy - The Amarna Eclipse - can now authenticate the story of King David, Solomon and Sheba!**

Thousands of ancient inscriptions scattered throughout the Arabian desert and especially Sheba point to Thamudic script as the original alphabet of letters. The explosion of alphabetic writing is centered around Midian, where we find the site of Mount Sinai. The original alphabet quickly split into a northern and a southern variety both of which are found at Sinai. As early as 1896, **Hubert Grimme, the famous historical linguist and scholar of Semitic languages, claimed Thamudic to be the original alphabet of letters. In 1923 Grimme**

**attributed the original alphabet to the Israelites of the Exodus.** Another well-known biblical archaeologist, Stephen Caiger, came to the same conclusion in 1936. Modern scholars now concede Thamudic is the oldest alphabet of letter symbols. Thamudic is not only the father script of the north Semitic alphabet (ancient Hebrew) but also of the south Semitic alphabet (Sabean, also called Minean).

## Grimme's Tales

In the late 1800's, the historical linguist Hubert Grimme traveled to the Mid-East to study Hebrew inscriptions. In 1896 he published his classic work on ancient Hebrew which included an appendix on *The Names of God* - which became more famous than the book itself. Grimme was one of the first to analyze the pictographic alphabet found at Serabit el-Khadim. In 1923, Grimme published "*Ancient Hebrew Inscriptions from Sinai.*"

> From the preceding explanation of the ancient Sinaitic inscriptions [it] can be concluded that their language is pure Hebrew which does not depart in any significant aspect from the language of the Bible, particularly not from the Bible's older parts, whereas it differs greatly from the other Semitic languages... such as Arabic, South Arabian and Aramaic.

Grimme was the first to study all of the inscriptions in the region and analyze alphabetic development. Grimme was clearly predisposed to the Bible narrative but his knowledge of Thamudic as well as the early inscriptions in the Proto-Semitic pictographic alphabet was impeccable. **Recent discoveries now support Grimme's findings** (see pp. 253-4). According to Egyptian records, during the long reign of Joseph's pharaoh Amenemhat III (1680-1633 BC), there were 38 expeditions from Egypt to Serabit el-Khadim. They would have been administered by his Vizier Joseph, undoubtedly using many Hebrew supervisors and workers, proving Grimme's conclusion "*The discovery that... Hebrew was spoken - and even written - on the Sinai Peninsula, revolutionizes completely the hitherto existing view of the history of the Hebrew language... one will have to attribute to them the leading role in transferring the Hebrew language* [and writing] *to the Phoenician homeland"* (Grimme 1923:78 & 91) [brackets mine].

> ...a rival arose against the Egyptian as well as the Babylonian writing systems, in the form of a writing system based on alphabetic

characters, which was destined to obliterate these earlier scripts… to a language in which pure Biblical Hebrew can be recognized, belongs the honor of having been first in using the new writing system… [for] the '*Ten Commandments*' and that the Semitic alphabet was the means of recording them.

According to Grimme... the Thamudene alphabet was not only the earliest North Arabian script, but it was also the Prototype of the Sabaean and the other South Arabian alphabets. He also considers the Thamudene alphabet as a direct offshoot of the Sinaitic script…

Grimme is practically alone in this opinion.
(Diringer, 1996: p.229)

Hubert Grimme was indeed alone in 1896 when he first proposed Thamudic script (Thamudene) was the original alphabet of letters. "*There is no doubt that the authors of the Bible kept in mind this Semitic writing system as the original written form of [the tablets of] the law*" (p.90). Grimme was a renowned scholar of ancient Hebrew and Semitic languages. We owe much to him for our knowledge of these ancient languages and how they evolved. Despite his profound expertise and experience in the field, Grimme's claim the Thamudic script was the original alphabet was rejected. More than a century later, "*scholars have no definitive answer to the problem of the earliest appearance of these inscriptions.*" (Naveh, 1982: p.43).

After all Sinai, meeting place of Egyptian and Semite, is, as Gardiner pointed out, a logical place for the development of the Semitic alphabet, especially since it is likewise geographically suited to be the point from which the two independent branches of Semitic script, North and South, spread. (Ullman, 1927:313)

Grimme did receive some support from scholars such as Ullman and Van Den Branden, but his theories were not accepted during his lifetime. In an article published in 1923, **Grimme stated Thamudic was the alphabetic writing of the Israelites!** Even though scholars offer no answers of their own to the contradictions of the Phoenician Origin Theory, most have been patronizing of the Exodus perspective, "*Nowadays these romantic views are no longer accepted*" (Naveh, 1982: p.26). In the past few decades, the latest archaeological and linguistic finds are confirming Grimme's conclusions. Thamudic is now considered by experts to be the oldest alphabet of letter symbols and its evolution from the Proto-Semitic pictographs is a given.

# Caiger & Grimme World's Earliest Alphabet!

In 1936, the prominent archaeologist Stephen Caiger arrived at the same conclusion as Grimme from a different direction. Caiger studied "*Minean*" script meaning the Thamudic alphabet that developed into the Sabean writing of south Arabia. The Sabean empire eventually broke up into four separate nations of which the Minean, to the north, became the most prominent. As a result, Sabean writing, and its predecessor Thamudic, was often referred to as Minean.

The Proto-Sinaitic pictographs had been discovered only a decade before and Caiger's research led him to conclude ***"the alphabet was invented* [at Sinai]*... thence carried to South Arabia: then to Canaan: and eventually to Phoenicia. ...far from questioning the ability of Moses to write, we may even conjecture that he had progressed…to an alphabetical style of writing...*"** (Caiger, 1936:p.5-6) [brackets mine]. This is the exact sequence of events posited by Grimme. We now know the alphabet born of glyphs in Sinai was turned into letters in Midian. Thamudic has been clearly established as the *Minean* (aka *Sabean*) script that Caiger focused on in his research.

> Recent discoveries indeed have shown that the alphabet is a far more ancient invention than was commonly supposed. The Minaean inscriptions of south Arabia, ... are dated by many scholars as early as 1400 BC, and exhibit well-developed alphabetic writing. Some are of the opinion that **the Hebrews of the pre-Conquest era 'must have originally employed this Minaean [Thamudic] script...**
> (Caiger, 1936: p.5) [brackets mine]

There is no doubt that Hubert Grimme's work deserves a detailed re-analysis in light of recent developments.

# Chapter Six

## Inscriptions at Serabit

*Joseph dwelt in Egypt, he and his father's house:*
*And Joseph lived one hundred and ten years.*
(Genesis 50:22)

The Bible says Joseph was 17 years old when taken to Egypt, spent 13 years in prison, worked under pharaoh during seven years of plenty, then in the second year of the famine brought Jacob and his family to Egypt beginning the sojourn in 1662 BC. This would make Joseph's entry into Egypt - 1683 BC in the first few years of the long (1680-1633 BC) reign of Amenemhat III according to the Canon of Turin.

The story of Joseph from the Egyptian record is fascinating, and includes not only the biblical account but also other major accomplishments only hinted at in Scripture. **Many of the inscriptions from Serabit el-Khadim are known to come from the reign of Amenemhat III. Joseph, his Vizier, would have been personally in charge of supervising those rich mines.**

In 2016, a modern historical linguist, Douglas Petrovich, took on the challenging task of translating the pictographic alphabetic inscriptions at Serabit el-Khadim. His conclusions have sent shock waves through the academic community. "***Hebrew has the distinction of being the world's first alphabet,*** *the one from which Phoenician and every other alphabetic script in the history of the world has derived.*" (Petrovich, 2016: p.192).

Douglas Petrovitch did the translation of these inscriptions. They fall into three categories. 1) An inscription could be personal but more likely 2) it related an important happening to the tribe, or 3) the nation of Egypt. Petrovich's translations of certain of the Amenemhat III inscriptions at Sinai are in the following pages.

The story we are relating of Joseph, the great Vizier, is fascinating, and includes not only the biblical account but also major accomplishments only hinted at in Scripture - but now ascertained by the Egyptian record. Many of the inscriptions from Serabit el-Khadim are known to come from the reign of Amenemhat III. These turquoise mines were a major source of wealth to the Egyptian throne. They would have been closely managed at the highest level by Joseph - Vizier of Pharaoh Amenemhat.

The mines at Serabit el-Khadim were in use over several centuries. During the long reign of Amenemhat III there were 38 expeditions to the turquoise mines of Serabit. Basically, that is an expedition every year during the winter when it was cool enough to work the mines.

**It is important to note that the historical context of these inscriptions is my own interpretation. Douglas Petrovich does not hold to the New Chronology.** So, there is no possibility his translations were biased in order to make them match the biblical chronology as exemplified in the New Chronology. This is highly significant since his translations fit perfectly anyway. I contacted him after publication of his book and he informed me his next book will be entitled: *New Evidence of Israelites in Egypt from Joseph to the Exodus.*

The inscriptions fall into three categories. An inscription could be personal (1), but more likely related an important happening to the tribe (2), or the nation [Egypt] (3). Petrovich's translations of the Amenemhat III inscriptions are as follows:

Sinai stele #115 - "*Levantines - Hebrews of Bethel - the Beloved.*" The engraving shows the arrival of a dignitary with followers. This inscription would likely refer to the arrival of Joseph, Jacob and his large tribe from the Levant [Canaan]. Bethel was the first community established in Canaan by Abraham (Gen 12:8).

The engraving may refer to the arrival of Joseph to Serabit el-

Khadim which would, to the Hebrew workmen, still be a reference to the arrival of the Levantine Hebrews in Egypt. To them Joseph was the most prominent Levantine Hebrew of them all. *Bethel*, in Hebrew, means the *house of El*, Yehovah the beloved.

Sinai stele #376 - "*The house of the vineyard of* ***<u>Asenath</u>*** *and its innermost room were engraved. They have come to life!*" If Joseph spent a significant amount of time at the isolated mines of Serabit el-Khadim, it is likely he would want his wife and sons with him. This inscription records the completion of a house and vineyard for ***Asenath***. In Genesis we are told that Joseph was given a wife, the daughter of the high priest, along with an Egyptian name, making him Egyptian royalty. Her name was ***Asenath***. "*And Pharaoh called Joseph's name Zaph-nath-paAneach, and he gave him to wife* ***Asenath****, the daughter of Potipherah priest of On*" (Gen 41:45).

Sinai stele #351 - "*After the year had changed, it* [the Nile] *was twice as swollen. Then the cultic shrines and the watering troughs were becoming unclean.*" The inscription is dominated by a huge image identified as Ptah. This is the inscription used for the cover of Petrovich's book below. Petrovich dates this inscription to the 18th dynasty. The time of Joseph's management of the mines was during the reign of Amenemhat III.

It is a little difficult to endorse Petrovich's conclusion here, as he himself says "*Ptah certainly was well known to the Hebrews who ventured to Serabit, because he was alluded to or depicted on stelae from the days of Hebeded's journeys to the mines during Dynasty 12, including Sinai* [inscription] *136 (Year 11), Sinai* [inscription] *92 (Year 13), and Sinai* [inscription] *100 (Year 20),* ***all from the reign of Amenemhat III***" (2016: p.126).

**That would be the time of Joseph. It is a dramatic confirmation of the Nile flood cresting at double its level - tamed by Amenemhat's great Vizier**. "*Given that the Egyptian calendar year begins with the inundation (of the Nile) season, and that the Nile River swelled at this time, the author undoubtedly was implying that the Nile is the entity that swelled to twice its normal level.*" (2016: p.125).

Sinai stele #349 -" He [pharaoh] *sought occasion to cut away to barrenness our great number, our swelling without measure, he performed terror against their quiver and brought about a cry of wailing"*[46] [brackets mine]. Exodus passages 1:7-22 relate the mandate from the pharaoh "*who did not know Joseph*" that the midwives destroy all male children of the Hebrews and that all Egyptians were charged with killing Hebrew male children by throwing them into the Nile.

Sinai stele #375a - "*The Overseer of Minerals* ***- Ahisamach*** *- the one having been elevated is weary to forget.*"[47] Now the Hebrews are enslaved, no longer workmen earning a wage, and *Ahisamach,* though put in charge, laments the new status of the Hebrews and his own curse in overseeing their oppression. During the Exodus (Exodus 31:6), **the task of cutting and setting minerals for the tabernacle is given to "*Oholiab, the son of Ahisamach*"** using skills taught to him by his father - Overseer of Minerals at the mines of Serabit el-Khadim.

Sinai stele #361 - "*Our bound servitude has lingered.* ***Moses then provoked astonishment!*** *It is a year of astonishment…*"[48] There is much more in Petrovich's translations relating to the context of the times of Joseph and the Exodus. We will conclude here because it ties us back into the Exodus narrative where *Moses* moves through the region sweeping up all the Hebrews under his banner. **Petrovich dates this inscription to dynasty 18, *circa* 1446 BC, the date of the Exodus.**

In *The Writing of God*, the Exodus story continues as told through the inscriptions at Mount Sinai in Arabia.[49] It is noteworthy that Petrovitich and myself came to the same conclusion - that Hebrew is the oldest alphabet of letters! **It would certainly seem that Hubert Grimme's prescient early work a century ago - has been validated!**

Many scholars have studied these inscriptions without much result. Petrovich's break-through came when he decided to identify the pictographic letters as Hebrew. **For those readers unfamiliar with decipherment of ancient inscriptions, if a string of letters make sense in a particular ancient language and none other, then you have found the master key.**

Why wasn't Hebrew tried earlier? The simple answer is that no one considered Hebrew as a possibility. The more complex answer is that many researchers are willing to contort history to insure that Hebrew is not a possibility.

> If one chooses to search at the wrong site, during the wrong period of history, and examine the wrong literary corpus, he/she is doomed to fail miserably from the very beginning. Essentially, this has created a 'straw man' argument that critical scholars have used to mock the historicity of the biblical account.
> (Petrovich, 2016: p.187)

This is a very courageous stance by Petrovich. Most researchers are loathe to make critical statements about the academy, even though - in this case - clearly true and easily documented. Publications, professorships, prestige and paychecks are all given from the hands of the established order which possesses the power to cast a scholar into outer darkness where their hard work will never be published, debated or even acknowledged.

"*Oddly enough, professional Egyptologists are not nearly as pessimistic about these matters as higher critical biblical scholars, whose unstated agenda is to create doubt about historical matters in the biblical text*" (Petrovich 2016: p.186). "*Hoffmeier (2015:205) informally attempted to poll 125 random scholars from the IAE's (International Association of Egyptologists) membership, from which 25 responded. When asked if they think the early Israelites lived in Egypt, and there was some sort of exodus, 19 of the 25 (76%) answered YES.*"

"*Moreover, none of the 25 answered NO to the question...* [which] *does reflect the existence of a substantial percentage of them who affirm the historicity of the Israelite sojourn-exodus account in the Bible... If biblical scholars were polled similarly it is difficult to imagine that 76% of them would support the historicity of the Israelite sojourn-exodus*" (Hoffmeier, 2015: p.205).

> When Thomas Levy held a large conference in 2013 on the topic of the exodus, no speaker presented positive evidence of the exodus or the sojourn, which is reflected in the same dearth of positive evidence in any of the 43 articles published in the follow-up volume of 2015. Even more astoundingly, none of the speakers/authors even commented

about 'personal research' they were conducting to determine whether or not such tangible evidence exists. If [no] evidence is being studied expressly to determine if an archaeological footprint of Hebrews in Egypt of the second millennium BC exists, what legitimate chance of discovering such evidence is there? (Petrovich, 2016: p.186-7)

***"The inescapable conclusion to draw is that there is no viable rationale as to why a massive conference/volume was formed around the topic of the exodus without even including actual research to determine if Hebrews were present in Egypt, especially during the time 'when the biblical text would require it' (i.e., 1876-1446 BC)"*** (Petrovich, 2016: p.187).

*"The historical credibility of the Israelite–exodus has been nothing but an object of criticism... the critical-scholarly world has cleverly built an impenetrable force field around the evidence, without any rational hope of breaching their defense, and not welcoming anyone into the academy who would challenge the status quo. Yet with the present work, a new day has dawned"* (2016: p.188). Let us pray Petrovich is correct about that "*new day*" dawning for the mainstream academy.

It is significant to note **both Douglas Petrovich and myself have come to the same conclusion working completely independently from two totally different corpuses of inscriptions.** I first published my work in 2010 based on the Thamudic inscriptions at Jebel al-Lawz, Mount Sinai in Arabia. Petrovich published his conclusions in 2016 based on the Proto-Semitic pictographic inscriptions from Serabit al-Khadim. Our research validates the previous study of these inscriptions by Hubert Grimme (1896) who studied the inscriptions at Serabit, and Stephen Caiger (1936) who studied Thamudic [Minean] inscriptions from the Arabian Peninsula. **Grimme and Caiger also came to the same conclusion - the original alphabet was Ancient Hebrew!**

# Chapter Seven

## The Sea Peoples

*Have I not taken up Israel out of the land of Egypt,*
*and the Philistines from Caphtor...*
(Amos 9:7)

The clues to solving the enigma of the date of the Exodus are found in the stars which outline the history of the Middle East. **Archaeo-Astronomy now reveals to us the true timeline and therefore the true story of the Exodus.** Possibly the most important, and certainly the most overlooked part of that story involved those Aegean tribes - called the Sea Peoples - who contested the territory of Canaan with the Israelites. Around 1000 BC the Sea Peoples were the ancestors of the Greeks who inhabited the islands and coasts of the Aegean Sea, called Achaeans.

The story of the Greek Sea Peoples merits close scrutiny. The Sea Peoples conquered the powerful Hittite empire, fought the Egyptians to a standstill, inhabited the coast of Canaan, engaged in a prolonged struggle with the Israelites for possession of the promised land then finally assimilated into the Aramean population in Phoenicia and Syria as well as settling in other parts of the Mediterranean. **The Sea Peoples most crucial to our story are the Philistines, their allies the Anakim, and the Kadmonites.** The Sea Peoples passed their seafaring knowledge on to the Phoenicians and took the alphabet back to Greece with them. The alphabet, the writing of God, was carried all over the Mediterranean via their seafaring and colonization. Even later, the Aramean language and alphabet, adopted from the Hebrew, would spread across the Middle East to India.

**We begin with a tale of two military campaigns into Canaan for two very different purposes, led by two very different Egyptian**

**pharaohs.** These two pharaohs had very similar names, **Shoshenk and Shishak**. The military campaign of Shoshenk I, also spelled Shoshenq, is part of the Egyptian record. Shishak is the pharaoh of the Bible who sacked Jerusalem. Equating these two has caused major havoc to history. **The New Chronology allows us to date them - resolving the confusion.**

Jean Francois Champollion (1790-1832)

**It was Champollion who made the initial error in confusing Pharaohs Shoshenk and Shishak**, a mess that continues to confound us to this day. Champollion is renown as the "*Father of Egyptology.*" It was he who cracked the code of the Rosetta Stone giving us the ability to read ancient hieroglyphs. Champollion opened the door to the understanding of the panorama of ancient history. This achievement rightfully evokes awe in those, like myself, who have studied his accomplishments.

In 1798 Napoleon led his army into Egypt to capture that crossroads of world trade - sitting as it did on the Mediterranean, giving access to Europe and the West, and the Red Sea giving access to Asia and the East. He brought with him a team of academics to study the ancient artifacts of Egypt. Napoleon's expedition was short-lived as the British soon attacked, destroyed his navy, and besieged his army trapping them in Egypt.

# The Rosetta Stone

While fortifying a fort in the town of Rosetta (Rashid) Egypt, in 1799, French Captain Pierre Bouchard discovered a large basalt monument written in three scripts and two languages: hieroglyphs, demotic [Egyptian cursive writing], and Greek. Although Egyptian had been a dead language for almost two millennia, ancient Greek was a well-known language. A public fervor spread through France, then Europe, that the key to understanding the secret mysteries of ancient Egypt would soon be known.

Intellectual giants threw their efforts into the competition to read the Egyptian glyphs. Sylvester De Sacy of France and Johan Akerblad of Sweden made a highly publicized attempt but, although they made some progress, ultimately failed. Some 15 years later Thomas Young, a renowned polymath, dedicated himself to the task. He determined that the hieroglyphs included phonetic signs and succeeded in deciphering the cartouche of Ptolemy V on the monument, but Young soon returned to his many other projects.

It was Champollion who became obsessed with the hieroglyphs and finally succeeded. Already possessed of skill in ancient Greek, Champollion learned Coptic. Coptic is the Egyptian language written in Greek letters by the Christians of Egypt called Copts. To Champollion it was the closest thing to ancient Egyptian, and his best option for deciphering the glyphs.

Taking the signs that Young had already deciphered, he was able to figure out the other phonetic signs among the hieroglyphs. With this framework and the bi-lingual inscription he was then able to determine which Egyptian signs represented the Greek words of the translation. In 1822, more than twenty years after their discovery, Champollion published his breakthrough on deciphering the Egyptian hieroglyphs of the Rosetta Stone.

In 1829 Champollion voyaged to Egypt to study the monuments in person. The journey up the Nile validated his ability to read the hieroglyphs and discern their meaning. The sheer number of inscriptions was overwhelming. They were everywhere! Champollion ranged up and down the Nile reading the cartouches of the pharaohs of antiquity with increasing awe and awareness. He soon realized the importance of establishing the order of succession of the pharaohs and their dates. **His overarching goal was to establish a connection between the Egyptians and the narrative of the ancient Hebrews in the Bible.**

As Champollion entered the temple complex at Karnak, more than 400 miles up the Nile, he found himself at long last standing before that connection. Along the Bubastite Portal, the gateway to the temple of Amun-Re, he identified the builder as Shoshenk I whose military campaign into Canaan was inscribed into the wall depicting impressive battle scenes and a list of cities conquered or subdued by the mighty pharaoh.

With rising exultation Champollion read off the Old Testament names of the cities of Canaan "*out from the pages of the Old Testament - Ajalon, Gibeon, Mahanaim, Rehob, Beit Shean, Shunem, Taanach, Megiddo. Then Champollion came across a name-ring that caused great excitement. He began to read the glyphs: Iuda ha Malkuth - 'Judah the Kingdom.' Here was a key link to the Bible" (Rohl, 2015:51).*

The story of Pharaoh Shishak's campaign against Canaan is told in the books of Kings and Chronicles. After the death of Solomon, his son now King "*Rehoboam dwelt in Jerusalem, and built cities for defense in Judah... Bethlehem, Etam, Tekoa, Bethzur, Shoco, Adullam, Gath, Mareshah, Ziph, Adoraim, Lachish, Azekah, Zorah, Aijalon, and Hebron, which are in Judah and Benjamin fenced cities*" (II Chronicles 11:5-10). "*And it came to pass, that in the fifth year of King Rehoboam, Shishak king of Egypt came up against Jerusalem... with twelve hundred chariots and threescore thousand horsemen... and he took the fenced cities which pertained to Judah... and took away the treasures of the house of Yehovah*" (II Chron 12:2-9).

Rehoboam - by bowing down to Shishak, surrendering Jerusalem and all its treasures - survived and continued to rule as a vassal of the Egyptians. **This then was the great and celebrated link validating the Bible as history and connecting it to the Egyptian record!** Unfortunately, historians connected it to the wrong Pharoah.

As I stated before - neither the tight embrace of the Bible - nor its rejection as history has served well the cause of truth. This is a perfect example. Champollion, in his eagerness to connect the Bible to the Egyptian record - got this wrong - and almost two hundred years later we are still trying to disentangle our view of history from the consequences of that error. Look at the list of cities conquered by Shoshenk, and contrast it with the list of cities fortified to protect Judah and its royal city of Jerusalem against Shishak. There is only one city, *Aijalon*, on both lists. Especially glaring is the lack of Jerusalem on the conquest list at Karnak. Well, what about '*Judah the Kingdom*?' By 1888, Wilhelm Muller had already proven Champollion's hasty translation of "*Iuda ha*

*Malkuth*" was incorrect. In fact, it reads "*Yad ha Melek*" (Hand of the King) an outpost near Megiddo.

## Shoshenk & Shishak

Since the biblical chronology cites the sacking of Jerusalem in 925 BC, the scholars who hopped aboard this train fixed the date of Pharaoh Shoshenk/Shishak at that time. Kenneth Kitchen, one of the most prominent Egyptologists alive today, and an evangelical Christian, calls this "*the essential synchronism*" for dating Egyptian history (Kitchen, 1987). The entire chronology of the New Kingdom is framed around this false anchor point. Shoshenk I, followed by his son Osorkon, was the first ruler of the 22nd dynasty, so the 20th and 21st dynasties with their 242 years of reigning pharaohs, are butted up to the reign of Shoshenk I in 945 BC, "*giving us an end-date for the 19th dynasty of circa 1186 BC. It is then just a matter of adding up the reign-lengths of the last five kings of the 19th dynasty, plus the 66 years of Ramesses II's long reign to arrive at an accession date for Ramesses the Great in circa 1279 BC*" (Rohl, 2015:59).

So, starting from an anachronism in Exodus 1:11 identifying the royal cities of "*Pi-Thom and Raamses,*" scholars have anointed Ramesses II (aka "*the Great*") as the pharaoh of the Exodus. Then seizing upon the Karnak inscription of Shoshenk's campaign into Canaan, Shoshenk is equated with Shishak of the Bible who sacked Jerusalem in 925 BC.

The backdating described above, seemingly justifies the later date of the Exodus as being in Ramesses reign 1279-1213 BC. **Thus, the Ramesses Exodus Theory is born and justified using Bible passages in a way that completely contradicts the biblical chronology of the Exodus.** There is nothing wrong with the methodology of the backdating. It is simply the false anchor point of equating Shoshenk with Shishak that is in error.

Number One, we already know when Shoshenk I reigned and it was not during the time of King Rehoboam. You may remember that the earliest fixed date in ancient history was 664 BC. It was the sack of Thebes in Egypt by the Assyrian King Ashurbanipal confirmed by so many synchronizations we need not recount them here. This date is accepted by all historians. The Bur-Sagale eclipse happened a century before that and was recorded by the Assyrians giving us a second fixed anchor. Thiele retro-calculated the Bur-Sagale eclipse to June 15, 763 BC during the ninth year of the reign of the Assyrian King Ashur-Dan III.

Using the Assyrian king list, which came from the cuneiform archives of Ashurbanipal's library, Thiele counted back through the kings on that list another century which brings us to the reign of Shalmaneser III from 858-824 BC. In the sixth year of his reign, he invaded Canaan fighting an alliance of kings including King Ahab of Israel (1st Kings), husband of Jezebel. So, from the date of that battle of Karkar recorded in sources outside the Bible and from the biblical chronology, Thiele ascertained Ahab's reign as 874-853 BC. Take the biblical chronology back several more kings and you have Jehoahaz of Judah who reigned 735-716 BC. The Assyrian king in his time was the great conqueror Tiglath-Pileser III, recorded in the Bible under his hypocoristicon of *Pul*. From the records of his campaign we have this:

> [**Tiglath-Pileser** received] the tribute of Kushtashpi of Kummuh, Urik of Que, **Sibittibaal of Byblos...** and **Jehoahaz of Judah**...

So, Jehoahaz' contemporary, just to the north in Byblos, was Sibittibaal (or Shipitbaal). A stone monument was found in Byblos with a Phoenician inscription which read, "*Wall built by* ***Shipitbaal, king of Byblos, son of Elibaal, king of Byblos*** *for Baalat-Gelel, his Lady. [May] Baalat-Gebel prolong the days of Shipitbaal and his years over Byblos.*" In Elibaal's reign he too made a dedication to the local goddess Baalat-Gebel on a statue of **Pharaoh Osorkon I**, "*statue which* ***Elibaal, king of Byblos, son of Yehimilk [king of Byblos]*** *made*

Pharaoh Osorkon I

*[for] BaalatGebel, his Lady. May Baalat-Gebel prolong [the days of] Elibaal and his years over [Byblos].”* Prior to Yehimilk in the reign of his father we have another of these customary statues of the reigning Pharaoh, overlord of Byblos, and dedicated to Baalat-Gebel. This final statue was of **Shoshenk I**. *“**On the Shoshenk statue,** found in 1894, the Phoenician text explicitly states that the gift from Egypt for the temple of Baalat-Gebel **was brought by King Abibaal himself.**”* (Rohl, 1995:370-371)

| | | |
|---|---|---|
| **Abibaal** | **Shoshenk I** | ***circa* 798 BC** |
| **Yehimilk** | **Shoshenk/Osorkon** | ***circa* 778 BC** |
| **Elibaal** | **Osorkon I** | ***circa* 758 BC** |
| **Shipitbaal** | **Tiglath-Pileser III** | ***circa* 738 BC** |
| **adapted from Rohl (1995:371)** | | |

**Note the methodology used to arrive at these dates. We started with two fixed anchor dates, then sought synchronizations (statues, inscriptions and war annals) between the king lists of Assyria, Byblos, Egypt and Judah to fix the dates of their king lists into the timeline. So the dates shown above are certain!** The margin of error would be within only a few years. There is far too much synchronism here to just dismiss or explain away. Solomon died in 930 BC and his son Rehoboam surrendered Jerusalem to Shishak in 925 BC. Pharaoh Shoshenk campaigned in Canaan more than a century later *circa* 798 BC.

## Shoshenk was not the biblical Shishak who sacked Jerusalem!

Point Number Two, there really was a Pharaoh Shishak who also marched into Canaan and we have Egyptian record of that campaign as well. ***Shisha***, or in Hebrew ***Shishak***, was the hypocoristicon of Ramesses II. In Egyptian, '*Ramesses'* is actually pronounced '*Riama-*

***Shisha***'. Thus, the hypocoristicon, *"which appears four times in the document known as Papyrus Anastasi I"* (Rohl, 2015:63). I have always considered it amusing that secular historians insist Ramesses the Great was the biblical pharaoh of the Exodus when, according to the Egyptian record, he campaigned in Canaan long after the Exodus, reigned for 66 years and died of old age in his bed. **To believe the secularists one must ignore both the Biblical and the Egyptian record!**

Thiele's date of 925 BC for the sack of Jerusalem is solid so that would put Ramesses II in the 10th century BC not the 13th. On the north pylon of the Ramesseum at Thebes, trumpeting the deeds of Ramesses II (the Great) we have this inscription:

**The town which the king [Ramesses II] plundered in year 8 - Shalem.** (from Rohl, 1995:149)

*Shalem*, or alternately *Salem*, from the root word *Shalom* meaning *peace*, was the original name given to *Yeru-shalaim* [Jerusalem], the /s/ and /sh/ sounds being somewhat interchangeable in the Semitic dialects of the era. In Genesis 14:18, "*Melchizedek king of Salem... the priest of the most high God"* blessed Abraham by the hand of Yehovah

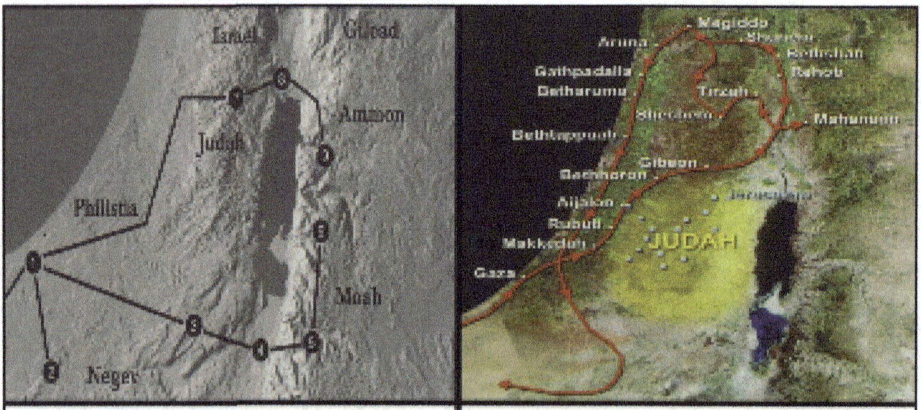


**Ramesses campaign 925 BC** | **Shoshenk campaign 798 BC**

Maps reprinted by permission of David Rohl.

**Although historians disagree with the New Chronology on the date of his reign there is no dispute that Ramesses II made a campaign into Canaan**. Kenneth Kitchen specifies the Ramesses campaign into Canaan during year 8 of his reign:

> Ramesses took things in hand. In no time he reached Gaza (1) … A flying column swept the Shoshu (Bedouin) back east, out of Canaan entirely (2). Then Ramesses dealt with East Palestine… the flying column struck down the Negev hills (3), across the Rift Valley south of the Dead Sea (4), and up into Edom-Seir (5) … swung north… into the heartland of Moab… to conquer Butartu (6)… At the same time, Ramesses himself swept round in a clockwise arc to complete the pincer movement - across the hilly central ridge of Canaan past Jerusalem (7) [which surrendered]... over the Jordan, past Jericho (8)… striking at Dibon (9). (Kitchen, 1982:67) [brackets mine]

The numbers at left give the cities taken by Ramesses during his campaign quoted above. He went into Judah to the south whereas Shoshenk invaded along the coast into the northern kingdom of Israel. **Aijalon was the only city entered by both.**

# Chapter Eight
## The Amarna Eclipse

***Yea, the stork in the heaven knows her appointed times... But my people know not the judgment of Yehovah!***
**(Jeremiah 8:7)**

Parallels are often drawn between the Greek Heroic Age and the warriors of King David and their exploits. They too were heroic warriors of that era, as we will see from a look at the **fascinating stories in the Amarna Tablets.** In 1887 an Egyptian woman discovered these clay tablets buried in ancient ruins in the desert outside her village. The 380 cuneiform tablets written in Akkadian - the *lingua franca* of the Amarna Age - were mostly correspondence between Pharaoh Akhenaton and his satraps ruling over the Levant [Canaan], giving us a window into Akhenaton's foreign affairs. In the Orthodox Chronology (OC), his reign is listed as 1351-1334 BC. **We now know this is not correct!**

**One of the Amarna tablets (designated KTU-1.78) describes an eclipse happening on the first day of the lunar month at sunset,** "*The day of the new moon of Hiyyaru was put to shame* [eclipsed] *as the sun (goddess) set, with Rashap as her gate-keeper.*" A complete solar eclipse is rare - an eclipse at any particular point on earth happens (on average) only once every 360 years. **The Amarna tablet recording an eclipse at sunset makes it easy to find. In 1988, Wayne Mitchell, using a mainframe computer at MIT, retro-calculated the eclipse at Ugarit in the month Hiyyaru [April/May] to May 9, 1012 BC at 6:09 pm. There was only one occurrence during this period so there can be no doubt as to the accuracy of the date.**

The mention of the gate-keeper Rashap is a fascinating confirmation as it has been well-established to refer to a giant star in the Crab

Nebula. Rashap was a commanding presence in the night sky - but this happened during the day - proof the sun was eclipsed so the giant star of Rashap could be seen clearly accompanying the sun into darkness. Two thousand years later Rashap exploded in a supernova, disappearing from our skies.

**The eclipse tablet, KTU-1.78, was found in the ruins of the palace of King Nikmaddu of Ugarit, burned and blackened by fire. In what David Rohl describes as "*one of those amazing pieces of fortune which so rarely occur in archaeology, we have the opportunity to date the burning down of the palace of King Nikmaddu.*"** In another of the Amarna letters the burning of the palace was reported to Akhenaton, "*To the king... Fire destroyed the palace at Ugarit... half of it has disappeared.*"[3] Another of the Amarna letters reconfirms his reign "*following the death of Amenhotep III, after Akhenaton had become senior monarch in his twelfth year, when Nikmaddu was ruler of Ugarit.*"[4]

So Pharaoh Akhenaton began his reign in 1012 BC, not 1351 BC as in the Orthodox Chronology. The astronomical retrocalculation of the Ugarit eclipse of 1012 BC matches the Amarna records and is in accord with the New Chronology and the biblical timeline - Akhenaton was contemporary with King David - who took the throne of Israel in 1010 BC. Therefore, the Amarna letters from Canaan, open a narrow window from 1020-980 BC - describe the biblical events of the Davidic period, which confirm our New Chronology timeline.

> El-Amarna letter EA 256 contains the names of five of leading players from the books of Samuel. The letter itself is written by Ishbaal, son of Saul, who in turn refers to Joab (David's military commander), Baanah (one of Israel's chieftains), David (ruler of Hebron and son-in-law of Saul), and finally Jesse (father of David).[5]

As reward for his victory over the giant Goliath, Saul had promised David the hand of his oldest daughter, Merab:

*So it came about at the time when Merab, Saul's daughter, should have been given to David, that she was given to Adriel the Meholathite for a wife. Now Michal, Saul's daughter, loved David. When they told Saul, the thing was agreeable to him. Saul thought, I will give her to him that she may become a snare to him, and that the hand of the Philistines may be against him.* (1 Samuel 18:19-21)

**Yehovah had made it clear to David that Jerusalem was to be his Holy City and David its king** (2nd Ch 6:6), but the Israelites had not yet been able to take it from the well-entrenched Jebusites. According to legend, young David placed Goliath's head on a spike before the walls of Jerusalem to give notice to the Jebusites. "*And David took the head of the Philistine, and brought it to Jerusalem*" (1st Sam 17:54).

At the Battle of Gilboa, the Philistines killed King Saul and three of his sons, including his eldest Jonathan. A letter to Akhenaton informed him that Ishbaal, Saul's remaining son, wrote a letter asking David to avenge the death of "*our*" father, thereby supporting the rule of Ishbaal; "*Wage war against the people of Gina* [the Philistines] *for having killed our father... Wage war against the king, your lord, as our father did, when he attacked Shunem, Burkuna, and Harabu, and deported the evil ones"*[6] [brackets mine]. These cities refer to the campaigns of King Saul, who is referred to in the Amarna letters by his Akkadian name, Labayu "*the Lion.*" David, having fled for his life from Saul, was now allied with the Philistines and in the pay of Akish, king of Gath. As Saul had planned and predicted, by wedding his daughter Michal to David, the other Philistine kings did not trust him. As a result, he was not asked to fight against Saul at Gilboa.

King Ishbaal (aka Ishbosheth) was soon assassinated by two of his own chieftains, Baanah and Rechab (II Sam 4:7). David as the sole remaining son (in-law) of Saul was crowned King of Israel (II Sam 5:1-5). King David played it cool, maintaining his alliance with the Philistines until he had solidified his control over the territory and army of Israel. David moved first against Jerusalem, held by the Jebusite King Abdi-heba.

With the rise of David's guerrilla army of *Habiru* - root of the word *Hebrew* - the letters of the Jebusite King Abdi-heba of Jerusalem to Akhenaton became more panic-stricken with every post: "*The land of the king is lost. All of it has attacked me. I am at war as far as the land of Seru (Seir) and as far as Gath... the Habiru have taken the very cities of* ***the king, my lord, all are lost... may the king, my lord, send his chariots to fetch me and my brothers... lest the Habiru kill us.***"

**During the Amarna Period, Pharaoh Akhenaton did nothing to fight back against the enemies of Egypt!** Akhenaton's fame came from his implementing monotheism in Egypt - in the form of the Sun God Aton - throwing out their pantheon of gods. Although there was a flowering of glory in art and architecture in the building of his capital city of Aketaton, the priestly elite soon brought the upstart pharaoh to his knees. He retreated to his capital where his embattled political situation and deteriorating mental condition left him helpless at the collapse of his empire. His religion of the one true god, Aton, was reviled.

Once crowned king, David returned to Jerusalem with his army. The Jebusites mocked him, saying - even "*the blind and the lame*" can keep you from coming into Jerusalem. "*Nevertheless, David took the stronghold of Zion: the same - the city of David*" (II Sam 5:7). David and his warriors went up a water tunnel to get inside the city walls and conquer the Jebusites. Then he turned against the Philistines to conquer their territory, leaving alone Gezer, Gaza, Ashkelon and Ashdod which were claimed by Egypt - not wanting to provoke Egypt to attack. In two major encounters with the Philistines (II Sam 5:17-25) in the valley of Rephaim he drives them from the hill country back into the coastal plains. Throughout his 40-year reign, King David never ceased fighting to secure and expand the kingdom of Israel.

King Solomon, David's son, also reigned for 40 years allying himself with Pharaoh Haremhab of Egypt, who offered him his second-oldest daughter to marry. Solomon accepted the deal which had advantages to both sides. Pharaoh now had an ally to protect and regulate trade routes through this crucial midpoint of commerce with the Greek Sea Peoples to the west, Hittites to the north and Babylonians to the east. Solomon had the protection of a military alliance with Egypt. Whereas King David had struggled without cease against Israel's enemies, the Bible portrays Solomon's reign as peaceful. King Solomon was able to turn his attention to commerce and grew rich as the broker of trade between the superpowers of the Mid East. Solomon purposefully had no children by Haremhab's daughter, moving her out of Jerusalem to her own palace on the road to Shechem. Solomon had no desire to be absorbed by Egypt via an '*Egyptian*' son sitting on the throne of Israel.

*And Solomon brought up the daughter of Pharaoh out of the city of David unto the house that he had built for her; for he said, my wife shall not dwell in the house of David king of Israel, because the places are holy, whereunto the ark of Yehovah has come.* (2 Chronicles 8:11)

The ruins of her Egyptian-style palace complete with Egyptian sarcophagi have been discovered to the north of the Old City's Damascus Gate. A queen worshiping foreign gods within the Holy City of Yehovah would have been an outrage.

**The artifacts found in and around the site attest to the possessions of the Egyptian queen of Israel,** "*a serpentine statuette, alabaster vases (late 18th Dynasty), a limestone funerary stela (depicting the deceased offering flowers to the god Seth), a hotep-style offering table; at least two palmiform column capitals, a scarab (bearing the motif of a khepri beetle used to adorn a funeral canopy), and a faience ushabti figurine.*" The tomb, identified by Gabriel Barkay, contains stone-cut sarcophagi, a type of burial coffin never utilized by the Israelites. The most critical discovery was that of a large heart-shaped scarab inscribed with the royal cartouche of King Seti, the pharaoh who succeeded Haremhab, which establishes the demise of Solomon's Egyptian queen around the mid-point of his reign.

**Keep in mind, none of this evidence fits into the Orthodox Chronology** and as a result, *"Today, many archaeologists working in the Levant reject the idea of an Israelite empire under its two great kings, David and Solomon... Solomon, if he ever existed, could have been nothing more than a local chieftain ruling over a rustic tribal court." - "There was no mass Exodus from Egypt. There was no violent conquest of Canaan. Most of the people who formed early Israel were local people."* **This is the outcome of denial of the evidence of the Exodus and errors in the conventional chronology - preventing all this evidence from being placed within its proper timeframe. The New Chronology has put the Bible back into the history business! Now my readers have the evidence to decide for themselves.**

# Chapter Nine

## The Star of Bethlehem

*Where is he that is born King of the Jews*
*For, we have seen his star in the east,*
*And are come to worship him.*
(Matthew 2:2)

The stars are fixed in the skies, rotating slowly to our view as the earth rotates on its axis. There were, however, stars that "*wandered*" through the sky - moving across the constellations - sometimes stopping, even going backwards. Today, we know these wandering stars to be planets. They are the brightest objects in the sky other than the sun and moon. It was these wandering stars the ancient astronomers followed with such diligence. The word *moedim* for '*appointed times*' is translated '*seasons,*' in Genesis 1:14, which obscures the meaning to the modern reader.

**Yehovah was, in fact, speaking to us from the night sky, telling the observer when things would occur.** "*Lift up your eyes and look to the heavens...*" (Isa 40:26). "*The heavens display the glory of God, day after day they pour forth speech, night after night they display knowledge... Their voice goes out into all the earth, their words to the end of the world*" (Psalms 19:1-4). The word *olam*, usually translated as '*world*' can also mean '*universe*' or '*forever*' giving the meaning of going throughout the universe even to the end of time.

**In this book we have been continually citing astronomical evidence of the dating of events in Scripture which validate their truth.** We discerned these facts from celestial events as "*they display knowledge*" thousands of years later, thanks to astronomical retro-calculation. This is not voodoo. It is, and always was, part of Yehovah's plan that we know of the "*appointed times*" of His prophecies! During the years of 3-2 BC the skies lit up like the Fourth of July with celestial events. When the planets intersect one another in close conjunction, they create a light in the sky brighter than the other stars. Jupiter, the largest planet, therefore the brightest star in the sky is called the king star, or king planet. It comes into close conjunction with another planet or bright star about every seven years. In one year, Jupiter became the brightest star in the heavens—coming into conjunction - seven times - about a 1 in 4 million probability!

Zachariah had a duty to serve at the Temple during the priestly course of Abijah which would have been in the eighth week of the spring and fall. In spring he stayed into the ninth week for the Feast of Shavuot. All priests were required to serve during the festivals since all Jews everywhere were required to be in Jerusalem for the three great feast days: Pesach (Passover), Shavuot (Pentecost), both in the spring, and Sukkot (Feast of Tabernacles) in the fall. **Zachariah was elected to serve in the Holy of Holies on Shavuot 20 May 3 BC.**

**On that day, the first of ten major conjunctions in the heavens occurred.** Mercury (the Messenger) conjoined with Saturn to form a great light in the sky. Zachariah was elected to represent the Jewish people in the Sanctuary. The angel Gabriel appeared to Zachariah and anointed him and his wife Elisheva, with the blessing of a child - Yohanan. Zachariah was struck dumb because of his disbelief. In the days following this event, although in advanced age, Elisheva conceived. Scripture says everyone in Judea knew of the events confirming Zachariah as the high priest of Israel by the archangel Gabriel (Luke 1:65).

Scriptural events during the appearance of the Star of Bethlehem:

1. The star was visible from Babylon, periodically, before Yeshua's birth to guide the Magi to the new king. (Mt 2:2)
2. Gabriel appeared to Zachariah in the Sanctuary. (Lk 1:11)
3. Zachariah and Elisheva conceived a son, Yohanan. (Lk 1:57)
4. The angel Gabriel appeared to Miriam when John was in his sixth month in the womb. (Lk 1:36) Miriam conceived a son by the Holy Spirit (Lk 1:35), to be called Yeshua. (Lk 1:31)
5. Yohanan is born, circumcised and named. (Lk 1:57-63)
6. An angel tells shepherds of Yeshua's birth. (Lk 2:8-18)
7. The Magi led by the Star in the heavens arrive in Jerusalem before King Herod seeking the king of the Jews. (Mt 2:1-2)

8. The Magi visit the child Yeshua in Bethlehem when the Star stops above the home of Joseph and Miriam. (Mt 2:9)
9. Joseph, warned by an angel that Herod wants to kill Yeshua, fled to Egypt with his family. (Mt 2:13)
10. An angel appeared in order to call Joseph out of Egypt, after Herodes Magnus (Herod the Great) died. (Mt 2:19-20)

## Celestial Events of Star of Bethlehem 3-1 BC

***Most events are in approximation to the dates given at left.**

Although it is possible, even probable, that Yehovah meant for the sequence of bright stars to mark other events - **the sole reason for the**

| Date | Conjunction | Birth Events of two Messiahs |
|---|---|---|
| May 20, 3 BC | Mercury & Saturn | Gabriel appears to Zachariah In the Sanctuary on Shavuot |
| June 13, 3 BC | Venus & Saturn | Zachariah and Elisheva conceive a son - Yohanan |
| Aug 12, 3 BC | Jupiter & Venus | A bright Star appears to the Magi in Babylon |
| Sept 14, 3 BC | Jupiter & Regulus in Leo | Gabriel appears to Miriam later she conceives Yeshua |
| Feb 17, 2 BC | Jupiter & Regulus in Leo | Yohanon is born to Zachariah and Elisheva |
| May 8, 2 BC | Jupiter & Regulus in Leo | The Magi begin their Journey to Jerusalem |
| June 17, 2 BC | Jupiter & Venus in Leo | At some point - early on - Yohanon taken into desert |
| Aug 26, 2 BC | Jupiter, Venus, Mars, Mercury | Foretells the birth of Yeshua the Messiah in Bethlehem |
| Dec 25, 2 BC | Jupiter stops in Virgo | The Magi visit the child Yeshua bearing gifts |
| Jan 10, 1 BC | Lunar Eclipse | Herod dies after the lunar eclipse & before Passover |
| Aug 20, 1 BC | Jupiter & Venus | An angel calls Joseph, Miriam and Yeshua out of Egypt |

* Chart adapted from Dourson, 2014: p.83, but I have made various changes.

**Bethlehem Star, given in Scripture, was to guide the Magi to worship the newborn king!** The entire sequence of bright stars may have been there only to guide the Magi. Even at the birth of Yeshua there is no mention of a bright star overhead. Angels appeared to the shepherds. I include the conjunctions in the chart because they coincide very closely to the scriptural events. They may either foreshadow them or announce them shortly after the event, either way they are in close conjunction to the events themselves. We will examine them - in order to see how close they may be.

I do not think these astronomical sightings are necessarily meant to give an exact date of the events. Nonetheless, it gives us a framework of the timeline of the birth of the two messiahs that has much to commend it. Since Elisheva was in her sixth month when Gabriel appeared to Miriam then we have a basic sequence of 15 months from the conception of Yohanon ben Zachariah to the birth of Yeshua Ha Mashiach.

Gabriel appeared to Zachariah on or about the week of Shavuot as given in Scripture since Zachariah's priestly duty was in the eighth week which was followed by Shavuot - one of the three great festival days that required all the faithful to come to Jerusalem for the feast. It is the only explanation why the "*whole multitude of the people were praying without at the time of incense*... (Luke 1:10). "*And the people waited for Zachariah, and marveled that he tarried so long in the temple*" (Luke 1:21).

We also have an anchor just past the end of this 15-month timeline, the death of King Herod. **Yeshua was born just before the death of Herod as it says in Matthew 2:1 and 2:19. Josephus recorded that Herod Magnus died between a lunar eclipse and Passover of that year.** Just before Herod entered the final days of his agonizing death, "*that very night there was an eclipse of the moon. But now Herod's distemper greatly increased upon him in a severe manner.*" Josephus was a first century recorder of these events. That makes it definitive! **The lunar eclipse occurred January 10, 1 BC which means that the following Passover was April 11, 1 BC. Herod Magnus died in 1 BC!**

For the past century it was thought Herod died in 4 BC, due to the writing of Emil Schurer in 1890. He estimated Herod's death from Josephus' statement, "*So Herod... died, having reigned thirty-four years, since he had caused Antigonus to be slain, and obtained his kingdom;*

*but thirty-seven years since he had been made king by the Romans.*" Schurer's estimate of 4 BC, based on 37 years of reign, pushing the date back three years, was rather shaky to begin with. **It does not match other historical accounts. However, it was buttressed by the occurrence of a partial lunar eclipse in that year.** "*This is traditionally ascribed to the eclipse of March 13, 4 BC. Unfortunately, this eclipse was visible only very late that night in Judea and was additionally a minor and only partial eclipse. There were no lunar eclipses visible in Judea thereafter until the year 1 BC.*"[10]

The months from May to September are less than six. "*Gabriel announced the conception of Jesus* [to Miriam] *'in the sixth month' of Elisheva's pregnancy (i.e., what we would call the fifth month).*"[10] In ancient times the primary way a woman knew she was pregnant was by a missed period. The conception could have happened anytime during a month-long cycle, for this reason, the calendar month of the missed menses was counted as the first full month. In this case that would have been May, or Sivan in the Hebrew calendar.

The Hebrews were on a lunar calendar with shorter months, 29-30 days each as opposed to our solar month of 30-31 days each. This backs up the date another week or more. Since we figure these dates based on the modern calendar - we count May, June, July, August - mid September was the end of the fifth month - the next day was in the sixth month! Elisheva's conception, following Shavuot, was in May on our calendar. So it is reasonable the visitation of Gabriel to Miriam was sometime around Sept 14, 3 BC. Shortly after that Miriam conceived by the Holy Spirit. Yohanan was conceived in May and born 9 months later, foreshadowed by the conjunction of Feb 17, 2 BC. **Six months after Yohanan's birth was the conjunction of Aug 26, 2 BC, announcing the forthcoming birth of Yeshua - some 15 months after the conception of Yohanon.**

# Chapter Ten

## The Coming of the Magi

*Seventy weeks are determined upon your people and upon your holy city... to anoint the most Holy... after threescore and two weeks shall Messiah be cut off.* (Daniel 9:24-26)

Another great step forward is the understanding of the story of the Magi. These Magi are described by Philo. Philo of Alexandria was the great philosopher of the first century who expressed Hebrew thought in Greek. He was contemporary with Yeshua and wrote of the Eastern school of Magi from Babylon. He also wrote about the divine perspective of Hebrew education which had been passed down and practiced since Sinai.

> Those... are born of heaven who are men of skill and science and devoted to learning; for the heavenly portion of us is our mind, and the mind of every one of those persons, who are born of heaven, studies the encyclical branches of education and every other art of every description, sharpening, and exercising, and practicing itself, and rendering itself acute in all those matters which are the objects of intellect. - Philo -

Philo's opinion of the Magi of the Eastern school in Babylon is clearly one of high praise, "*Among the Persians there is the body of the Magi, who, investigating the works of nature for the purpose of becoming acquainted with the truth, do at their leisure become initiated themselves and initiate others in the divine virtues by very clear explanations.*"[2] These Magi were descended from those trained by Daniel. Daniel's prophetic visions of the future were so real to him that he was sometimes weak and pale for days afterwards. Yehovah sent the angel Gabriel to interpret them for him. Apparently, Gabriel was charged with bringing the prophecy about. **The essential part of Daniel's prophecy was the 70-week ministry of the Messiah. It was to begin after Herod rebuilt Jerusalem, "*the mandate to restore and to rebuild Jerusalem… And after threescore and two weeks shall Messiah be cut off*"** (Dan 9:25-26). Yeshua the Messiah was cut off, crucified after 62 weeks, two score and two weeks.

Michael Rood was the first to link Daniel's 70-week prophecy to the ministry of Yeshua. I refer you to his research in *The Chronological Gospels*. The reasons these Magi were initiates of Daniel are several. Daniel stayed in Babylon devoting himself to learning, teaching and serving Yehovah. Daniel trained his followers. That these Magi, followers of Daniel's teaching, were Hebrews is obvious! Why else would they care about a prophecy of the birth of the king of the Jews?

Specifically, Daniel trained them to watch for the signs of the coming of the Messiah… "*going forth from the commandment to restore and rebuild Jerusalem*" (Dan 9:25). It was Herod who "*restored and rebuilt Jerusalem*" - her Temple, fortresses, roads, bridges, aqueducts. Herod the Great was named such - because he was possibly the greatest builder of antiquity! When the Magi heard of Herod doing a massive rebuilding of Jerusalem they began to watch the skies closely. Only one scriptural event can be correlated with any certainty to the Star of Bethlehem!

Miriam would have gone up to the Temple in Jerusalem to sanctify the birth of Yeshua, after 33 days of ritual purification. I presume they would have brought Yeshua to the Temple for this ritual at the Feast of Sukkot in October 2 BC, soon after the Magi arrived. The Magi had long known that Jupiter - the brightest star in the sky, called the '*King Star'* - was the Star of the Messiah. After a long sojourn in Jerusalem, probably awaiting the next appearance of the Star, the Magi departed. In late December of 2 BC, Jupiter appeared directly in front of them, due south, leading to Bethlehem. When they found the home of Joseph, Miriam, and Yeshua - Jupiter stopped over it!

**On Dec 25, Jupiter began its retrograde motion, stopping and then turning backwards across the sky. Due south over the home of Joseph and Myriam - the Star stopped low on the horizon!** "*The star which they had seen in the east* [in Babylon], *went before them, till it came and stood over where the young child was. When they saw the* *star, they rejoiced with exceeding great joy. And when they were come into the house, they saw the young child with Miriam... and fell down and worshiped him"* (Matt 2:9-11). **It was the first '*Christmas*' as the Magi celebrated the newborn king with gifts of gold, frankincense and myrrh. Retro-calculation reveals this was Dec. 25, 2 BC.**

After the Magi's visit on Dec 25, 2 BC, an angel warned Joseph that Herod sought to kill Yeshua and they fled to Egypt in midwinter. Herod died early in 1 BC. His son, Herod Archelaus took over Judea and, fearing a revolt, pre-emptively slaughtered 6000 Pharisees! The slaughter provoked an actual revolt. General Varus had to bring his legion from Syria to quell the revolt killing 2000 more Jews. Archelaus was eventually banished to Gaul, not for his ruthlessness - but for monumental incompetence. By the end of the summer the fighting was over. Aug 20, 1 BC was the last conjunction of Jupiter with Venus, occurring about the time the angel came to call Joseph home (Matt 2:20), fulfilling the prophecy "*When Israel was a youth I loved him, and out of Egypt I called my son*" (Hosea 11:1).

It would be nice to pinpoint, with assurance, the exact date of Yeshua's birth. It seems quite possible we are not meant to know. That day is not to be known beforehand. "*But of that day and hour knows no man, no, not the angels in heaven, but my Father only*" (Matt 24:36). There exists a belief circulating for millennia that the Messiah will return on the same day he was first born.

In discussion of the book, '*Sons of Zion vs the Sons of Greece*' there is no greater contrast than that between the pure and innocent baby Yeshua, the Prince of Peace; and King Herod - once a great builder - now a bloodthirsty madman! "*Herod, when he saw that he was mocked of the wise men, was exceedingly wroth, and went forth, and slew all the [male] children that were in Bethlehem, and in all the countryside thereof, from two years old and under*" (Matthew 2:16). Power was the identifying virtue of Hellenism. Ruling was the proof of their superiority.

**To the Grecians, like the Herods and the Romans, the will to rule meant a complete insensitivity to the suffering of others. The '*noble*' Greeks were proponents of ruthless violence!**

# Philo cites the difference between Hebrews and Greeks:

> Moreover, it is ordained in the laws themselves that no one shall do to his neighbor what he would be unwilling to have done to himself... That a man shall not take up what he has not put down, neither out of a garden, nor out of a wine-press, nor out of a threshing-floor; and that absolutely no one shall take anything ...whether it be great or small, out of a heap. That no one shall refuse fire to one who begs it of him. That no one shall cut off a stream of water, but that everyone shall contribute food to beggars and cripples, and that each shall have favor with God.

Pressure on the Hebrews to convert to the Grecians' pagan gods and Hellenized ways led to revolt in 68-70 AD. The Romans brought in their legions and destroyed the Temple, killing hundreds of thousands. In 135 AD the Jews revolted again. This time the Romans razed Jerusalem to the ground slaughtering all her people, completing the catastrophic aftermath predicted after the 70 weeks prophecy of Daniel (9:26). "*The people of the prince that shall come, shall destroy the city and the sanctuary.*"

Even as his death was approaching, Herod continued killing. He burned alive two beloved rabbis, because their students destroyed the Roman standard, a bronze image of an eagle, which Herod had set up at the entrance to the Temple. Just days before his death, Herod executed his son Antipater. It was because of this madman that the angel warned Joseph to flee to Egypt with Yeshua. It was also why Yohanan was taken into the desert shortly after his birth, out of the reach of Herod and his equally ruthless sons. "*The child grew, and waxed strong in spirit, and was in the deserts till the day of his shewing unto Israel*" (Luke 1:80). Zachariah, warned, passed the blessing of the priestly line onto his son, *Yohanan Ha Matbil - John the Immerser* - known to us as John the Baptist who, via that blessing, was in turn, the true high priest of Israel.

# The Murder of Zachariah by Herod Magnus:

***The Protoevangelion of James* details the murder of Zachariah by Herod. It is often called the *Infancy Gospel of James the Just*, the**

**brother of Yeshua.** Scholars suggest it was written in Hebrew in the 2nd century AD by followers of Yohanan and Yeshua living in Antioch. It would have been passed down by James and recopied or retranslated in the 2nd century. It is a recounting of the events of the conception and birth of Yohanan and Yeshua we have been discussing. Perfectly consistent with the accounts of Matthew and Luke - it also details other aspects of the story, such as the killing of Zachariah.

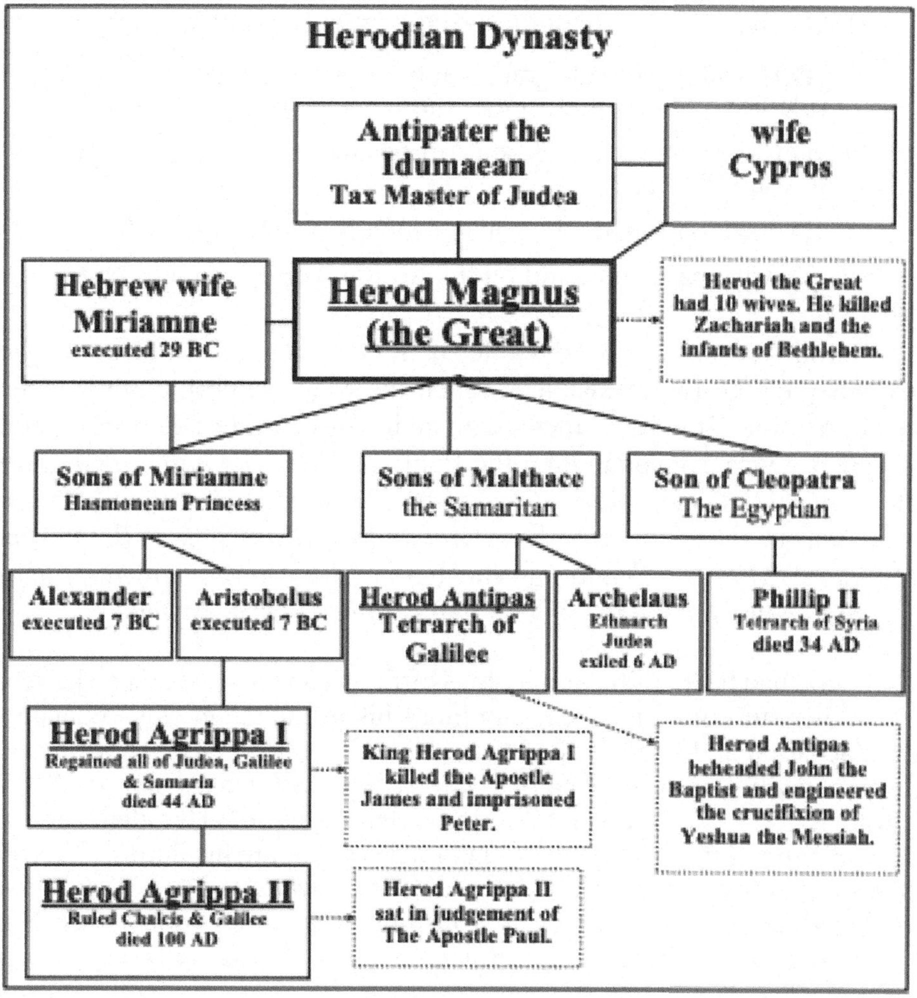

# Timeline of the Herodian Dynasty:

**37 BC** Herod Magnus made King of Judea, later Galilee & Perea.

**2 BC** Yeshua Ha Mashiach born in Bethlehem.

**1 BC** Herod Magnus (the Great) died, divided his kingdom.

**1 BC** Herod Antipas becomes Tetrarch of Galilee and Perea.

**30 AD** Herod Antipas had John the Baptist beheaded.

**31 AD** Herod Antipas engineered the crucifixion of Yeshua.

**39 AD** Herod Agrippa I, grandson of Herod Magnus, ruled over a reunited Judea, Galilee and Samaria.

**39 AD** Caligula ordered Herod Antipas banished to Gaul for trying to usurp Agrippa.

**44 AD** Agrippa I died, His son Agrippa II ruled over Chalcis.

**59 AD** Agrippa II met Paul on his way to Rome for judgment.

Herod was not knowledgeable of Scripture and did not know the prophecy of the two messiahs. When the Magi came looking for the newborn king of the Jews, they assumed it would be Herod's son. Herod assumed it was Yohanan (John the Baptist), "*when he had gathered all the chief priests and scribes of the people together, he demanded of them where the Messiah should be born. And they said unto him, in Bethlehem of Judea" (Matt 2:4-5)* quoting Micah 5:2, "*But you Bethlehem... out of you shall come forth unto me one that is to be ruler in Israel.*"

Herod had heard of the miraculous birth of Yohanon. So, when he sent his soldiers to kill all the male children - his prime target was Yohanan - but Zachariah and Elisheva had been warned. "*And when Elisheva heard that Yohanan was being sought she took him and headed for the hills*" (Proto 22:5). The soldiers did not find Yohanan, but the Herodian agents, sent *incognito* into the Temple, found Zachariah doing his duties. "*Herod continued seeking Yohanan and he sent his servants to Zachariah at the altar, saying to him, 'Where did you hide your son?' But he replied, saying to them, 'I am here as a servant of Yehovah and am serving in his temple. How should I know where my son is?' And his servants departed and reported to him all these things. Outraged, Herod said, 'Is his son destined to rule Israel?'*" (Proto-Evangelion of James, 23:1-5).

**Herod sent his secret police again and they questioned Zachariah again and brought back the same answer. Finally, "*Around daybreak, Zachariah was murdered*,** *even though the children of Israel did not know...*" (Proto 23:9). When the priests investigated his disappearance in the Sanctuary, they found no corpse, only his blood. The priests ripped their robes and wept, then "*they went out and reported to the people that Zachariah had been murdered. And all the tribes of the people heard, and they mourned and wept for three days and three nights*" (24:1011). After three days they cast lots for whom should take the place of Zachariah, "*And the lot went to Simeon. For he was the one to whom it had been revealed by the Holy Spirit that he would not see death until he saw the Messiah in the flesh*"[5] (24:13-14). **When Herod learned of Yeshua - the Messiah went immediately to the top of Herod's kill list!**

It was the death of Zachariah, (Yeshua's cousin-in-law, more like an uncle to him), whom Yeshua was referring to in Luke (11:50-51) when he said, "*the blood of all the prophets, shed since the foundation of the world, may be charged against this generation, from the blood of Abel to the blood of Zachariah, who was killed between the altar and the sanctuary.*" Was Zachariah, father of Yohanon, a prophet? "*And his father **Zachariah was filled with the Holy Spirit and prophesied,** saying, Blessed be the Lord God of Israel...as he spoke by the mouth of his holy prophets, which have been since the world began*" (Luke 1:6770). **Then the blood of Zachariah is counted in "*the blood of all the prophets shed since the foundation of the world*."**

Yeshua knew of the murder of Zachariah, "*From the blood of Abel unto the blood of Zachariah, who perished between the altar and the sanctuary... For you have taken away the key of knowledge; you yourselves did not enter, and you hindered those who were entering*" (Luke11:51-2). This generation of priests condoned a political takeover of the Temple that was corrupt to the core. The Temple was one of the wonders of the ancient world, but the heart of it was evil. The Jewish leaders exchanged substance for show, a show designed to benefit Herod, his priests and the Romans, not Yehovah. For that reason, the Temple, 46 years in the making, was to be destroyed by the hand of God.

In Matthew 23:35 it adds "*Zechariah, son of Berechiah, whom you murdered between the Sanctuary and the altar.*" That addition is not in the very same quotation in Luke 11:51. This presents a problem. The

Church deemed their Scripture inerrant - so the interpretation of the Church could not be challenged. My position is that the word of God is indeed inerrant - but only God's truth is inerrant, not man's truth! This error is different, it appears to be a simple scribal error. The addition of "*son of Berechiah*" is not in Hebrew Matthew. In *The Hebrew Gospels*, Zachariah is called a "*blessed son*," which in Hebrew (*ben barooch*) looks almost identical to (*ben berech*, son of Berechiah). Yeshua was making a profound condemnation of the Hebrew priests by saying they had persecuted and killed all the messengers Yehovah had sent them from the beginning of humankind until the present. Backdating that statement by six hundred years erases Yeshua's profound message.

Zachariah, father of Yohanon, and Zechariah are the same name. I am using the different spellings to distinguish between the two prophets. The variant spelling '*Zacharias*' is the Greek ending in /s/, rather than the Hebrew ending in /h/. It is a common change to many Hebrew names in the Bible due to the dominance of the Greek translation. Zechariah was a prophet of the 6th cent. BC. We do not have any detail in Scripture how Zechariah, son of Berechiah, died. There is yet another Zechariah:

> *Then the spirit of Yehovah took possession of Zechariah son of the priest Jehoiada... 'Thus says Yehovah: why do you transgress the commandments of Yehovah, so that you cannot prosper? Because you have forsaken Yehovah, he has also forsaken you.' But they conspired against him, and by command of the king they stoned him to death in the court of the house of Yehovah.* (2 Chronicles 24:20-1)

However, this Zechariah, son of Jehoiada, lived in the 9th century BC. Zechariah (Zech 1:1), the son of Berechiah, died in the 6th century BC! There is no evidence in Scripture that this Zechariah, son of Berechiah, was killed "*between the altar and the sanctuary*." So Zechariah, son of Berechiah, could not be the priest cited in Matt 23:35! To fix this problem, it is proposed that Zechariah, son of Berechiah, was the last of the Old Testament prophets. So Yeshua's charge of '*all the prophets whose blood was shed*' (Luke 11:50) has been interpreted as referring to only those in the Tanakh, ending six centuries ago. **That is not only wrong - also unthinkable to a Messianic of the first century with the Temple still standing, prophecy still in the wind, and the fresh blood of Zachariah still wet on the Temple floor!**

To deny Zachariah, father of Yohanon, was murdered by Herod

would also display little, or no, knowledge of the reality of Judea in the first century AD. The Herods were ultimately responsible for all the deaths in Judea of the followers of the two messiahs until the end of their dynasty in 44 AD - including Zachariah, Yohanon, Yeshua, the Apostle James, Stephen - the Nazarenes (followers of Yeshua) and Ebionites (followers of Yohanan) rounded up and killed. Acting under the authority and control of Herod - the chief priests purged Judea of all those who did not accept Herod's right to rule and the high priest's authority. Their power was inseparable from that of Herod!

The Herods, even after their father's death, continued to search for the Messiah, apparently unaware there were two - but acutely aware the Messiah was the rightful ruler of Israel according to Scripture, verified by miracles at birth and supported by the people. The two messiahs' popularity with the people also made them a target for Herod Antipas' deadly machinations.

Yohanan and Yeshua followed the law in all things so it is certain they began their priesthood at the age of 30 years old as required in Scripture. It says in Luke 3:23, "*Yeshua began to be <u>about</u> thirty years of age.*" In *The Hebrew Gospels* [HGC] there is no word '*about.*' "*Yeshua was beginning thirty years of age.*" Yohanan began some six months earlier than Yeshua. "*In the 15th year of the reign of Tiberius Caesar... the word of God came to Yohanan, the son of Zachariah, living in the wilderness, and he came into all the country about Jordan, preaching the baptism of repentance for the remission of sins... the voice of one crying in the wilderness*" (Luke 3:1-4) to prepare the way.

> "*Yeshua came from Galilee to Jordan unto Yohanan, to be baptized of him... and Yeshua when he was baptized... the heavens were opened unto him, and he saw the Holy Spirit descend upon him like a dove and lighting upon him: And a voice from heaven saying, 'This is my beloved son in whom I am well pleased.*" (Matt 3:13-17).

**Yeshua was anointed king of Israel by the only person authorized to do so, Yohanan the true high priest of Israel,** *Cohen Ha Gadol*, anointed by the angel Gabriel through his father Zachariah. When the Pharisees were asked, "*What was the source of Yohanon's baptism? Was it from heaven or from men?*" (Matthew 21:25). They would not answer.

The beginning of the ministries of the two messiahs started with Luke's statement, "*In the 15th year of the reign of Tiberius Caesar...*" I refer you to Chris Jones' excellent work, *Mashiach and the Sign of Yonah,* for the details of that research.[8] "*Knowing Pilate's appointment occurred during Tiberius' twelfth year of reign in 26 CE, it's thereby known Tiberius' fifteenth year of reign - when the Word of Elohim came to Yochanan - occurred in 29 CE.*" Yohanan's ministry began mid-year, and the baptism of Yeshua on, or about, Feb 15, 30 AD.

## Herod Plots Against the Messiah

**"*Beware of the leaven of the Pharisees and Herod*" (Mark 8:15). "*The leaven of the Pharisees, which is hypocrisy*" (Luke 12:1).** "Then Yeshua said unto them, '*Take heed and beware of the leaven of the Pharisees and Sadducees*" (Matt 16:6). The message of Yeshua was clear and consistent, Herod is a hypocrite who will use deceit and death to achieve his evil ends and so will his sycophants - the Pharisees and Sadducees (Matt 23:1-35). What seems to have been lost is the recognition that the Herods were behind all of these deaths and no such action would have been taken against either Messiah, their families or followers without Herod approving, indeed, directing it!

So why the circumspection in the Gospels about the Herods' ultimate responsibility for the killing of Zachariah, Yohanan, Yeshua and the others? Number one, it is clear enough from the Gospel record that the Herods were pulling the strings. Things that were known like the slaughter of the children at Bethlehem or the beheading of Yohanan were okay to write, without commentary. Actions Herod was trying to avoid responsibility for, like the crucifixion of Yeshua, were not okay to write about. Number two, the Herods and the Romans ruled. Any accusation or condemnation by believers against them would result in the immediate slaughter of the movement. To have or to hold to a Gospel that contained such accusations would be a death sentence. The early Hebrew writers had to be circumspect and the latter (Greek) ones had lost the Hebrew perspective and did not really know. Yeshua knew who was behind it all! He spoke to Pilate, while being condemned to death, saying, "*He that delivered me unto you has the greater sin*" (John 19:11).

"*Herod, the tetrarch, being reproved by him* [Yohanan] *for* [marrying] *Herodias his brother Phillip's wife, and for all the evils*

*Herod had done... so that he shut up Yohanan in prison*" (Luke 3:19-20) [brackets mine]. Herod Antipas intended to hold a show trial to discredit, condemn and execute Yohanan. Still, he feared the rebellion of the people if he killed him.

"*On Herod's birthday, however, the daughter of Herodias danced before them and pleased Herod. Whereupon* ***he promised with an oath to give her whatsoever she would ask. And she, ...*** *being before instructed by her mother, said 'Give me here John the Baptist's head... And he sent, and beheaded John in the prison*" (Matthew 14:6-10). Thus did Herod Antipas kill Yohanan.

Yeshua's first encounter with the Pharisees came when he healed a man with a withered hand on the Sabbath day in their synagogue. The Pharisees were unable to contest with him over the law. "*And he said to the man with the withered hand, 'Stand forth'. And he said unto them* [Pharisees], *'Is it lawful to do good on the Sabbath days, or to do evil? To save life, or to kill?' But they held their peace.*" Yeshua was aggrieved and angry at the hardness of their hearts. Then Yeshua said to the man, "*Stretch forth your hand. And he stretched it out and his hand was restored whole as the other.*" (Mark 3:3-5). ***"And the Pharisees went forth, and straightway took counsel with the Herodians against him, how they might destroy him"*** (Mark 3:6). The first thing they did was to seek approval of their plot from their masters, the Herodians, how they might destroy Yeshua.

"*At that time Herod the Tetrarch heard of the fame of Yeshua. And said unto his servants, 'This is Yohanan* [John the Baptist], *he is risen from the dead*" (Matthew 14:1-2). The plotting of the Pharisees went straight to Herod whose blessing and counsel they sought in order to destroy Yeshua. This is the first Herod became aware there were, in fact, two messiahs. It was certain that Herod would seek Yeshua's death. Now Herod must find a pretext to make the Romans do it, since only they had the authority over capital punishment.

Previously, Herod Archelaus had been exiled to Gaul, making it clear provoking the people to revolt would not be tolerated by the Romans who had to send legions to quell the outbreak. "*One of the most terrible in Jewish history... and this not without the loss of several thousand Roman troops.*"

Judea was taken away from the Herods and the Sanhedrin stripped of authority to impose the death penalty. In 26 AD, Pontius Pilate was made governor of Judea. Herod Antipas was Tetrarch of Galilee and Perea, though he still had a palace in Jerusalem. **Loss of territory and authority to execute his subjects, created hatred between Herod Antipas and Pilate - but for Herod to destroy Yeshua, he must reconcile with Pilate.**

At first the priests sent their disciples, "*Then the Pharisees plotted how they might entangle him in his talk. And they sent out to him their disciples along with the Herodians*" (Matt 22:15-16). "*so they might deliver him unto the power and authority of the governor*" (Luke 20:20). "*Master, we know that you say and... teach the way of God truly: Is it lawful for us to give tribute unto Caesar, or not? But he perceived their deceit and said... 'Show me a coin. Whose image and inscription does it have?' They answered and said, 'Caesar's.' And he said unto them, 'Render unto Caesar, therefore, the things which are Caesar's, and unto God the things which are God's*" (Luke 20:21-25). They could not contest with him and held their peace.

However, the Pharisees were not all of the same cloth, some had previously warned him about Herod, "*At that very hour, some Pharisees came to Yeshua and told him, 'Leave this place and get away, because Herod wants to kill you*'" (Luke 13:31). When the select session of the Sanhedrin was called to condemn Yeshua, only the reliable conspirators on the ruling council were included. Many among the Pharisees believed that Yeshua was a prophet. Joseph of Arimathea is one such example of a believer among the Pharisees of the Sanhedrin.

**Josephus wrote "*there was a 'political' Sanhedrin of Sadducee collaborators with Roman rule.*"** Their priestly order had taken over the leadership of the Temple during the latter days of the Hasmonean dynasty, once it became Hellenized and corrupt itself. Herod kept the Sadducees in place because they were reliably corrupt, pliable and useful. **Yeshua often referred to them as *Sopherim*, '*Scribes*,' calling them "*a brood of vipers.*"[11]** The high priest was always chosen from among the Sadducees. Herod the Great was a devoted Hellenizer and had ultimate power over the Temple and the Sanhedrin. Many Pharisees were also convinced, or coerced, or corrupted into following him:

> In the early days of Herod the Great... Menahem, possibly the Essene

of that name of whom Josephus tells us that he prophesied Herod's future greatness. The latter was tempted by the king's growing power, and, with eighty followers, entered into his service, forsook the ranks of the Pharisees, and appeared in gorgeous apparel, glittering with gold.

The culmination of Yeshua's ministry was symbolized by his triumphant entry into Jerusalem on Palm Sunday for Passover. "*And when he was come into Jerusalem, all the city was moved, saying, 'Who is this?' And the multitude said, 'This is Yeshua the prophet of Nazareth of Galilee*'" (Matt 21:10-11). The people spread their garments before him and strewed palm branches in his path and, cried "*Hosanna to the son of David: Blessed is he that comes in the name of Yehovah, Hosanna in the highest!*" (Matt 21:8-9) as Yeshua entered riding upon a donkey and a colt "*that it might be fulfilled what was spoken by the prophet, 'Behold your king comes unto you, meek, and sitting upon an ass, and a colt the foal of an ass*'" (Zechariah 9:9).

Yeshua threw the money changers out of the Temple once again, "*And the blind and lame came to him in the Temple; and he healed them. And when the chief priests and scribes saw the wonderful things, he did, and the children crying in the Temple, and saying, Hosanna to the son of David; they were sore displeased*" (Matthew 21:13-15). "*It is written, my house shall be called the house of prayer, but you have made it a den of thieves*!" There in Jerusalem was to be the final showdown!

The next day "*when he came into the Temple, the chief priests and the elders... came unto him as he was teaching, and said, 'By what authority do you do these things? And who gave you this authority?*" And Yeshua responded, "*I will also ask you one thing... if you tell me... I will likewise tell you by what authority I do these things.* ***The baptism of Yohanan, where did it come from? From heaven, or of men?***" The chief priests were stunned knowing if they said it came of God, he would ask why they did not believe him and obey. "*But if we say of men, we fear the people; for all hold Yohanan as a prophet.*" So, the chief priests said they could not say, and Yeshua responded, "*Neither shall I tell you by what authority I do these things*" (Mt 21:23-7).

"*That same day came to him the Sadducees, which say that there is no resurrection...*" posing a trick question about the wife of a man with seven brothers, each of whom died and each in turn took the brother's wife to marry, to whom would she be married in heaven. Yeshua replied, "*You err, not knowing the Scriptures, nor the power of Yehovah. For in the resurrection... they neither marry, nor are given in marriage, but are as the angels of God in heaven*" (Mt 22:23-30).

"*But when the Pharisees had heard that he had put the Sadducees to silence, they gathered together*" (Matt 22:34). And after they questioned him, Yeshua asked of them, "*What do you think of the Messiah? Whose son, is he? They said unto him, the son of David. He said unto them, How then does David in spirit call him Lord, saying, 'The LORD said unto my Lord, Sit on my right side, until I make of your enemies your footstool.' 'If David calls him Lord, how is he his son?' And no man was able to answer him a word, neither did any man... ask him any more questions*" (Matthew 22:42-46 quoting Psalm 110:1).

**The Messiah had confounded all the chief priests, Pharisees and Sadducees putting them to shame. As he was healing the blind and the lame, the priests were plotting to kill him. As Yeshua was fulfilling the prophecies of the coming of the Messiah, these Herodian priests were seeking to discredit the will and the Word of Yehovah.** In Yeshua's teaching in the Temple that day he excoriates the Herodian priests (Matthew 23), and by implication their master Herod, with a sustained ferocity scarcely matched anywhere in the Bible!

Yeshua said they put heavy burdens of observance upon the faithful but not upon themselves. Their works are done to be seen and evoke an image of righteousness. They revel in the egotistical privileges of their rank, the prominent seats, and greetings of respect by others, but do not humble themselves to serve. They neither enter into the kingdom of heaven nor do they lead others to it. In fact, they prevent the faithful from entering and their very prayers are for the purpose of self-adoration.

The greater their hypocrisy, the greater their damnation shall be. They make false oaths but seem to know nothing of the important matters of the commandments; judgment, mercy, and faith. They are naught but a sepulcher, gleaming white on the outside but full of rot and disease inside. They say they would not have taken the blood of the prophets **- but Yeshua says they are the measure of their fathers! They are vipers and cannot escape the damnation of hell (Matthew 23:1-33).**

"*Wherefore, behold, I send unto you prophets, and wise men and scribes: and some of them you shall kill and crucify; and some of them you shall scourge in your synagogues, and persecute them from city to city*" (Matt 23:33-34). "*You serpents, you brood of vipers, how can you escape the damnation of hell?*" **After this judgement of condemnation, the die was cast!** "*Although they wanted to arrest him, they feared the crowds, because they considered Yeshua a prophet*" (Matt 21:46). "*Then assembled together the chief priests, and the scribes, and the elders of the people, unto the palace of the high priest, who was called Caiaphas, and plotted how they might take Yeshua by subtlety and kill him*" (Matt 26:3-4).

It was Caiaphas who had previously argued for Yeshua's death when he raised Lazarus from the dead after four days in the grave. "*Then gathered the chief priests and Pharisees a council, and said,* ***'What do we do? For this man does many miracles. If we let him alone, all will believe in him'****... And one of them, named Caiaphas, being the high priest that same year, said unto them... 'Now consider that it is expedient for us, that one man should die for the people, so that the whole nation does not perish... he declared that Yeshua should die*... ***Then from that day forth they plotted together to put him to death*"** (John 11:47-53).

Yeshua could no longer walk among the people without the danger of the Herodian priests trying to seize him. Yet still he went to the Passover. "*Many of the Jews therefore knew that he was there: and they came not for Yeshua's sake alone, but that they might see Lazarus also, whom he had raised from the dead. But the chief priests plotted that they might put Lazarus to death as well; because by reason of him many of the Jews went away, and believed in Yeshua*" (John 12:9-11).

During the Passover week, Yeshua was seized by Temple guards and servants from the Herodian chief priests late at night in the Garden of Gethsemane. Yeshua was taken bound unto Annas the previous high priest and current president of the Sanhedrin. The Romans had appointed him in 6 AD when they took direct control of Judea from Archelaus. Annas was deposed in 15 AD for "*imposing and executing capital sentences which had been forbidden by the imperial government.*"

Annas remained the power behind the priestly throne, first through his son-in-law Caiaphas [who was also his nephew] who became high priest, and then by using his five sons, who all became high priests in turn. After Caiaphas came Ananias, who convinced the Romans to throw the Apostle Shaul [Paul] in prison and engineered the stoning death of James the Just [Yakov Ha Tzadek], brother of Yeshua, the leader of the Jerusalem Church.

After Annas, Yeshua was sent to the palace of Caiaphas, the current high priest, who convened a kangaroo court of '*select*' members of the Sanhedrin, "*And with him were assembled all the chief priests and the elders and the scribes*" (Mark 14:53). These '*scribes and elders*' were the Sadducees who would vote to condemn the Messiah to his death. They brought false witnesses against him but their testimonies conflicted. Then Caiaphas asked him, "*Are you the Messiah, the son of Yehovah*?" Yeshua responded, "*I am*" (Mark 14:61-2), whereupon Caiaphas rent his garments and declared no need of further witnesses. "*You have heard the blasphemy!*" All present condemned Yeshua to death for blasphemy without further ado, a judgment forbidden by Rome, which they had no authority to impose. They therefore "*plotted*" to find a way to make Pilate impose their illegal sentence of death. "*All the chief priests and elders of the people plotted against Yeshua to put him to death*" (Matt 27:1) The plot, in which Herod was to play a major part, was a clever one.

"*And when they had bound him, they led him away, and delivered him to Pontius Pilate, the governor*" (Matt 27:2). Even though Yeshua answered nothing to the accusations of the chief priests, Pilate was reluctant to crucify him and he offered to release a prisoner, either Yeshua or the murderer Barrabas, "*But the chief priests and elders persuaded the multitude that they should ask Barrabas, and destroy Yeshua*" (Matt 27:20). Pilate scourged Yeshua and brought him forth, "*when the chief priests and officers saw him, they cried out... crucify him*!" (John 19:6).

This was all part of the plot, the crowd was stacked with those officers and members of the Sanhedrin who had been seeking his death for some time, as well as agents of Herod's secret police to coerce the other Jews present. **Herod and his priests knew they had to put on a show to get Pontius Pilate to take responsibility for crucifying the Messiah.**

And put on a show - they did! Pilate had taken the measure of

Yeshua and did not want to put him to death. He discovered Yeshua was Galilean and "*as soon as he knew that he belonged unto Herod's jurisdiction, he sent him to Herod, who himself also was at Jerusalem at that time*" (Luke 23:7). Herod was reportedly delighted, wanting to see a miracle (Luke 23:8). One might imagine his delight as that of a cat playing with his prey. Yeshua's own words condemn Herod as a hypocrite, full of lies and deceit, his hands red with blood. Herod's delight was akin to that which he displayed toward Yohanan the Baptist (Mark 6:20) - before he cut off his head as a tip for a '*lap dance*' by Salome. Yeshua refused to be toyed with and said nothing. Herod and his soldiers mocked him and sent him back to Pilate.

Still, Pilate was reluctant. Blasphemy meant nothing to the Roman governor, '*go and judge him by your laws,*' Pilate said. The charge had to be sedition, someone claiming to be '*king of the Jews*' who would lead the Jews to revolt while they were gathered together for Passover. "*If you let this man go, you are not Caesar's friend: whosoever makes himself a king speaks against Caesar. When Pilate heard this, he brought Yeshua forth and sat down in judgement... Pilate said unto them, Shall I crucify your King? The chief priests answered, 'We have no king but Caesar!'" (Jn 19:12-15).* This crowd was not of the common people, nor even the Pharisees, who hated Roman rule and would choke on the phrase, "*We have no king but Caesar!*" These are Herodians intent on condemning Yeshua!

Pilate seemed convinced by this display that the Jewish people wanted him dead, as he argued with Yeshua, "*Your own nation and the chief priests have delivered you unto me: What have you done?*" (John 18:35). Then Herod comes to Pilate:

> *And the same day Pilate and Herod were made friends together: for before there had been enmity between themselves. And Pilate, when he had called together the chief priests and the rulers and the people, said unto them, You have brought this man unto me... and behold, I, having examined him before you, have found no fault in this man...*
> *No, nor yet Herod: for I sent you to him; and, lo, nothing worthy of death is done by him.* (Luke 23:12-15)

Herod's charade of friendship toward Pilate in order to manipulate him is easily proven. If Herod had truly vindicated Yeshua then the chief priests would have desisted in their demands to execute him. Herod was not one to be crossed if you valued your life. Instead, "*They were instant with loud voices, requiring that he might be crucified. And the voices of them and of the chief priests prevailed. And Pilate gave sentence that it should be as they required*" (Luke 23:23-4). "*Then he delivered him... to be crucified*" (John 19:16). This was all done so that Yeshua, the Passover lamb, be declared without blemish, as was required by Scripture.

Herod knew that he had Pilate trapped. The Roman governor could not ignore sedition since they had convinced him all the leaders, and the populace, were insistent Yeshua must be crucified. What if they complained to Caesar? What if the people did rise in revolt as they had done so frequently in the past? Pilate might be deposed, or worse! The charade had worked. Herod Antipas, the engineer of Yeshua's crucifixion, was held blameless. Pilate had already, during his tenure, executed other wannabe messiahs. Rome had a simple, brutal and bloody solution for all those who threatened revolt… crucifixion!

"*When Pilate saw that he could not prevail against them, but that a tumult was made, he took water and washed his hands before the multitude, saying, 'I am innocent of the blood of this just person...' Then answered all the people, and said, 'his blood be on us and on our children*'" (Matthew 27:24-25). This was a powerful curse for those present to call down upon themselves! I have my doubts whether even the most Hellenized and hypocritical of the Sadducees would say this. Most likely it came from Herod's thugs sent to fill out the crowd.

The Herods were famous for their secret police - often called upon to manifest a hyper-aggressive mob! In any case, this was not the voice of the Jewish people, since they believed Yeshua to be a prophet, and the Herodians feared a riot if he was seized, "*not during the feast... or there may be a riot among the people*" (Mark 14:2). The apostles, in Luke 24:20, were in no doubt who was responsible for their Messiah's death, "*The chief priests and our rulers delivered him to be condemned to death, and have crucified him.*"

**Yeshua Ha Maschiach was crucified on Wednesday April 25, 31 AD.**

It was the preparation day; Passover began at sundown of that day. Having gone through a lot of research, and archao-astronomy, to establish Yeshua's birth date - the year of his death follows as a natural consequence. Neither are we alone in this conclusion, just as the vast majority of early church fathers and writers agreed on the 2 BC birth year, so did many early historians cite 31 AD as the year of the Messiah's crucifixion. These include; Epiphanius of Salamis, Eusebius of Caesarea, Maximus the Confessor, and Georgius Cedrenus.

Yeshua's death can also be pinpointed from the work of Aurelius Cassiodorus, who wrote that the "*crucifixion occurred in the same year that Sejanus became Consul of Emperor Tiberius,*" which was in 31 AD. In addition, Josephus cited "*Pilate's appointment occurred during Tiberius' twelfth year of reign in 26 CE.*" Thus, as we have done previously, it is possible to extrapolate from Luke's timeline (Luke 3:1) "*Now in the fifteenth year of the reign of Tiberius Caesar...*" that Yeshua's baptism and ministry began early in 30 AD, ending in his death early next year, at *Pesach* [Passover], April 25, 31 AD. Messiah's birth & death:

| **Early Fathers - birth 2 BC** | |
|---|---|
| Irenaeus of Lyons | 2/3 BC |
| Clement of Alexandria | 2 or 3 |
| Eusebius of Caesarea | 2 or 3 |
| Epiphanius of Salamis | 2 or 3 |
| Hippolytus of Rome | 2 or 3 |
| Hippolytus Alexandria | 2 or 3 |
| Origen Adamantius | 2 or 3 |
| Quintus Tertullian | 2 or 3 |
| Sextus Africanus | 2 or 3 |
| Paulus Orosius | 2 |

| **Early Fathers - death 31 AD** | |
|---|---|
| Aurelius Cassiodorus | 31 AD |
| Maximus Confessor | 31 |
| Eusebius of Caesarea | 31 |
| Epiphanius of Salamis | 31 |
| Georgius Cedrenus | 31 |
| | |
| | |
| | |
| | |
| | |

# Breaking of the Lintel in *The Hebrew Gospels*

In three instances Jerome preserves evidence of a tradition in the early church of the breaking of the Temple lintel stone rather than the Temple curtain at the crucifixion of Jesus (Matt. 27:51; Mark 15:38; Luke 23:45). "*In the Gospel we often mention, we read that the immense temple lintel fell and broke to pieces.*"[41] Again, ***"In the Gospel, however, which is composed in Hebrew, we read not that the Temple curtain was torn but that the Temple lintel of wondrous size fell."***

In *The Hebrew Gospels from Catalonia* [HGC], it is the breaking of the stones of the Temple that are cited, which would crack in two the lintel stone of the Holy of Holies. Compare the verses given in the King James Version [KJV] to the HGC:

**HGC Matt 27:51** ***"And here the Temple was broken on both sides up and down and the earth quaked and the stones split in the middle."***

KJV Matthew 27:51 "*And, behold, the veil of the temple was rent in twain from the top to the bottom; and the earth did quake, and the rocks rent.*"

**HGC Mark 15:38** ***"The temple was broken on both sides up and down..."***

KJV Mark 15:38 "*And the veil of the temple was rent in twain from the top to the bottom.*"

**HGC Luke 23:45** ***"And the sun went dark and covered the temple and it was split in the middle."***

KJV Luke 23:45 "*And the sun was darkened, and the veil of the temple was rent in the midst.*"

The earthquake would account for the breaking of the large lintel stone over the entrance to the inner Temple, "*the stones split in the middle.*"

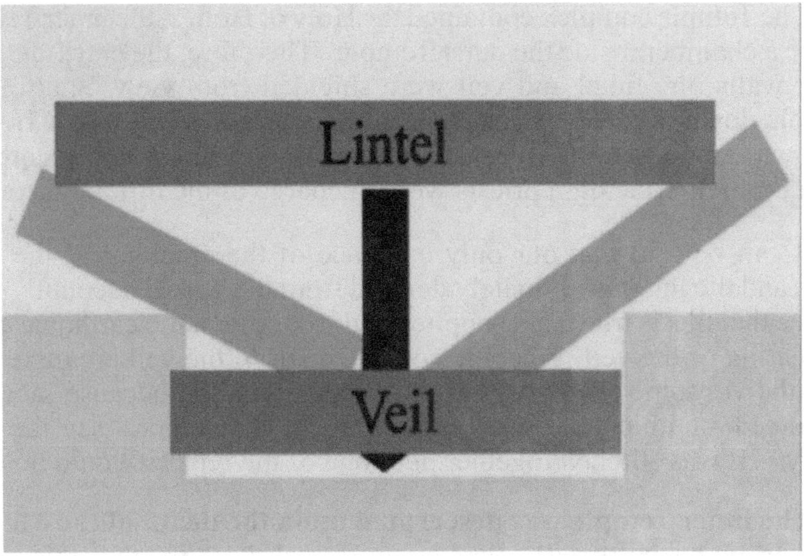


Note that the damage would also have torn the veil hanging beneath the lintel. However, the veil alone could hardly be ripped by an earthquake absent other damage. **The early church tradition of the breaking of the lintel - as preserved by Jerome from *The Hebrew Gospel* - is confirmed in all three verses of the HGC... Now we have a pattern!**

So, the pagan critics of Christianity were soon denying the events surrounding the crucifixion. Another group denying these events were the high priests of Herod's Temple. After all, they had everything to lose. Their magnificent Temple was a cash cow! The post of the high priest had been on sale to the highest bidder for almost a century now - since the time of the dissolute and corrupt descendants of the Maccabees.

The Herodians were even worse. Sacrifice of animals '*without blemish*' was required of every congregant on a regular basis. Upon arriving at the Temple, the Sadducee priest would inform the petitioner that their animal was blemished, but… it would be accepted as a payment and for an additional fee an animal without blemish would be provided. So, the priests gained the goat and the price of a goat. For their tithes, the faithful could not offer money with a pagan image, including the image of Caesar, but… for a fee the money changers could provide coin without pagan imagery. The Temple created immense wealth for Herod and the corrupt high priests, serving by his favor and authority.

The Temple complex contained the Holy of Holies, the inner Temple, inside a chamber inside the outer Temple. Therefore, the cracking of the inner walls, the lintel and veil were shielded from view. Shutting the Temple down because of this desecration was not an option. Covering it up was an option. Keeping all mention out of the official record was not a problem - the high priests were in charge of the official record!

It can be said that our only evidence of the cracking of the inner walls and the lintel is ultimately derived from the gospel account. To the degree that this is true it also applies to the darkness, the earthquake, and the tearing of the veil. Accounts of the repair of the veil are mixed and speculative, some saying it must have been repaired because sacrifices continued - unthinkable without the veil. Other reports say the huge heavy veil was stitched together but none of the repairs would hold.

**The inner Temple was desecrated upon the death of the Messiah - never to be restored!** Personally, I feel strongly that this potent symbolism matches the facts that we know from both Gospels. At the same time, I acknowledge that historical reports of these events outside of Scripture - except for one - can be challenged as based upon speculation, hearsay or imagination.

Shortly before Yeshua's death on the cross, onlookers including the chief priests, mocked him saying, "*you that would destroy the temple and rebuild it in three days, save yourself*" (Matt 27:40). Yeshua dies soon after, the earth quakes and the inner Temple is split from one side to the other, from the top to the bottom... breaking in half the huge lintel over the entry to the Holy of Holies which tears the veil beneath it in two. The most apt symbolism is the prophesied total destruction of the Temple, foretold by Yeshua (Luke 21:6), and Daniel (9:26), which came to pass by the Romans in 70 AD only 40 years later!

The veil beneath the lintel at the entrance of the Holy of Holies kept believers from entering the abode of Yehovah. Symbolically, the splitting of the veil removed that obstacle for anyone, especially Gentiles, communing with Yehovah. One can see how this symbolism would be important to the Greek Gentile Church. To Messianic believers the prophetic destruction of the Temple, foreshadowed by the breaking of the large lintel over the Holy of Holies - and the cracking of the inner Temple itself - was by far the most significant event! **It is not surprising then that *The Hebrew Gospels* would focus on the breaking of the lintel while *The Greek Gospels* focused on the tearing of the veil.**

Matthew wrote *The Hebrew Gospel* less than ten years after the fact, long before the first Greek converts or Greek Gospels.

The breaking of the lintel is a huge claim to make - both for historical and religious reasons. What evidence do we have this actually happened? The same question applies to the other phenomena mentioned - the earthquake, the tearing of the veil, the splitting of stones and the darkness that covered the Temple.

"*Thanks to Thallus, who wrote only twenty years after the events themselves... the 'darkness' that occurred was a historical fact that secular writers felt a need to try to explain away in some way.*" Jerome was not the only writer to mention these events.

> In the *Gospel of the Nazarenes* we read that at the time of Messiah's death the lintel of the Temple, of immense size, had split. Josephus says the same and adds that overhead awful voices were heard which said: '*Let us depart from this abode!*'

Critics attack the testimony of *The Hebrew Gospels from Catalonia*, calling it a '*corrupted*' text, saying there is no historical evidence of the cracking of the walls of the inner Temple and the lintel. Never mind that same criterion would negate the '*tearing of the veil*' in *The Greek Gospels* as well - for which there is no historical evidence outside of Scripture either. After all, an earthquake that could crack the walls and lintel of the Holy of Holies and rend the veil would have damaged other structures. This brings us to the one report we know to be historically valid.

> According to the Talmud, it states; Shabbat 15a -
> "*Forty years before the destruction of the Jerusalem Temple, the Sanhedrin was removed - from the Chamber of Hewn Stones in the Temple - and convened in the Market Place*"

> "*There is every reason to believe... that the Chamber of Hewn Stones was so damaged in the same earthquake that it became structurally unsafe from that time forward. Something like this had to have happened because the Sanhedrin would not have left this majestic chamber (to take up residence in the insignificant 'Trading Place') unless something like this explanation took place.*" The proof of this is that the Sanhedrin never returned to their majestic Council Chamber!

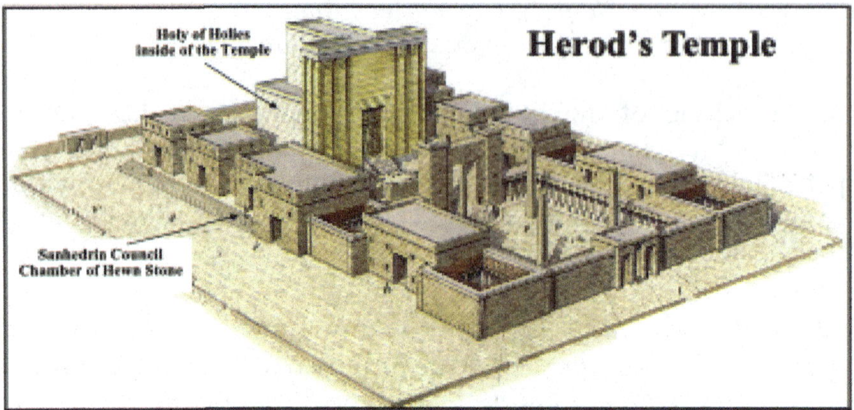


**The outer walls of the Temple concealed the damage to the Holy of Holies,**

The Temple was destroyed on the 9th of Av 70 AD by Titus - later to become Caesar - in putting down a rebellion of the Jews against Roman rule. Forty years before the destruction of the Jerusalem Temple was the most momentous year in history. The month of *Av* (*Nissan* in the new calendar) is the month of Passover. That year, from Passover 30 AD to Passover 31 AD, encompassed Yeshua the Messiah's one year ministry and his crucifixion. "*The Talmud tells us that when the Sanhedrin ceased to judge capital offenses, they moved from the... Chamber of Hewn Stones to the 'Market Place.'*"

> It was the very year of the crucifixion that the Jews were denied the right to perform capital punishment by the Romans. When the members of the Jewish Supreme Court [the Sanhedrin] brought Jesus to Pilate, he told them, "*Take him and judge him by your law.*" But they replied, "*It is not lawful for us to put any man to death*" - (John 18:31). Yet they connived and pressured Pilate and stirred up the crowd to demand the crucifixion of Christ the Messiah, the Anointed One of God. (John 18:32-40).

According to Luke 22:66, the daybreak trial where Yeshua was condemned to death for blasphemy - for claiming to be the "*son of God*" - took place in the Council Chamber at the Temple. According to the record, ***"the judgment made by the official Sanhedrin against Jesus within the Chamber of Hewn Stones, was the last judgement ever given by the official Sanhedrin in their majestic chambers within the Temple!"***

**Yehovah Himself gave testimony to Yeshua as the "*son of God*"** by covering the Temple in darkness, causing an earthquake to split the stones - cracking the walls of the inner Temple, splitting the lintel, rending the veil in two - desecrating the Holy of Holies forevermore and rendering the Sanhedrin Chamber unusable! Readers must decide whether there is enough evidence to accept the testimony of Yehovah as given in the Hebrew Gospels and the Greek Gospels. Their witnesses do not contradict - but rather confirm each other - each giving more detail. As for the Sanhedrin, from that day forward they deliberated over the law in the Market Place - accompanied by the bleating of sheep and the smell of manure!

# Chapter Eleven

## Three Days and Three Nights

***So shall the Son of Man be three days and three nights in the heart of the earth.***
**(Matthew 12:40)**

Much of the difference between the Hebrew tradition of Scripture and the Greek tradition is a matter of interpretation rather than changes in the text. Nonetheless, differing doctrine is often the driving force behind changes introduced by scribes and translators. Preserving the canon of the Gospels verbatim is more of a modern concept. Scribes of the era sometimes did that well, but at other times considered it their duty to correct errors in the manuscript. Certainly, those scribes doing the major compilations, such as the Greek edition of Eusebius, the Latin Vulgate of Jerome, and the Hebrew Masoretic text... were all focused on fixing errors, cleaning up the texts and perhaps uprooting any incorrect wording. It seems safe to say your perception of the '*correct wording*' might be different if you were a Greek Christian translating Hebrew Matthew into Greek - or a Messianic Jew translating the Greek letters of Paul into Hebrew.

The tradition of celebrating '*Holy Week*' - '*Good Friday to Easter Sunday*' is the most contentious example of differing interpretation of the Gospel Scriptures. The tradition is based on **the prophecy of the "*sign of Yonah*" (Jonah) made by Yeshua.** Of all the prophecies which Yeshua made certain to fulfill, it was the most repeated (3 times), the only prophet Yeshua associates himself by name (Matthew 12:41), and the sole prophecy upon which Yeshua declared his authenticity may be judged:

> *An evil and adulterous generation seeks after a sign; and there shall be no sign given to it but the sign of the prophet Yonah: for as Yonah was three days and three nights in the great fish's belly; so shall the Son of Man be three days and three nights in the heart of the earth.* (Matthew 12:39-40)

"*There is no wiggle room in this prophecy.*" As stated, it means three complete days and three complete nights. The biblical standard (Deut. 13) for any prophecy is rather severe. 1) It must have a written scriptural precedent. 2) The sign must come to pass exactly as stated. 3) If it proves false the '*prophet*' is rejected and put to death!

In the previous chapter we went to a lot of effort to ascertain that the date of Yeshua's crucifixion was Wednesday, April 25, 31 AD. This was the day of *Pesach* [Passover] in 31 AD. It was the preparation day. At dusk the Passover festival was to begin. The Messiah was entombed at dusk that day in the tomb of Joseph of Arimathea, "*It was the preparation day, and the Sabbath* [Passover] *was beginning*" (Luke 23:54). Mark agrees, "*Now it was already evening, and since it was preparation day...*" (Mark 15:42). Joseph had about 30 minutes of twilight in which to get the body placed in the nearby tomb, wrapped in linen, and the rock rolled shut before Pesach began.

The problem of interpretation arises with the use of the phrase, "*on the third day he will be raised to life*" as it is said in Matthew 17:23 and 20:19; Luke 18:33 and 24:7, and Acts 10:40. Yeshua himself even says it in Luke 24:46, "*This is what is written: The Messiah will suffer and rise from the dead on the third day...*" In English this could mean 1) Friday, 2) Saturday, 3) Sunday - the third day - but it has already been stated by Yeshua as three full days and three full nights. Does that create a contradiction in the Gospels? No, not necessarily, if and only if Yeshua arises at the end of the third complete day, and there have been three complete nights - as the chart below illustrates:

| Passover 31 AD | Wednesday April 25 | Thursday | Friday | Sabbath | Sunday |
|---|---|---|---|---|---|
| NIGHT | | | | | |
| DAY | | | | | |

The '*Good Friday - Easter Sunday*' scenario assumes, for one thing, that Pesach was on a Friday. It was in 33 AD - but not in 31 AD. Even on that date, it wrongly assumes that Friday and Sunday can be counted as partial days. Both Matthew (28:1) and Mark (16:2) are clear. "*At the end of the Sabbath, as it began to dawn, toward the first day of the week [Sunday], came Mary Magdalene and the other Mary...*" the stone was already rolled away and the angel informed them that Yeshua was already risen and gone. Mark reads the same, "*And very early in the morning* ***the first day of the week, they came unto the sepulcher at the rising of the sun...*"** **and Yeshua Ha Mashiach was already gone!**

The entombment at dusk on Friday was already past the end of the day. So that cannot count as a partial day. Yeshua was risen and gone before dawn according to the angel. So, Sunday cannot be counted as a partial day. It can be argued that it is already Sunday by both Roman, and Greek, (at midnight) reckoning. However, neither interpretation fulfills the prophecy of the sign of Yonah. Even if one concedes the Greco-Roman reckoning and allows counting the two partial days it still does not fulfill the prophecy. There are still only two complete nights! Here is the *Good Friday - Easter Sunday* scenario:

| Passover 33 AD* | Wednesday | Thursday | Friday April 3 | Sabbath | Easter Sunday |
|---|---|---|---|---|---|
| **NIGHT** | | | | | |
| **DAY** | | | X | | X |

**As Chris Jones so succinctly puts it in his analysis, "*It's a disparity with eternal consequences and ought to pierce every believer to their soul...*** *To theorize Yahshua invoked the sign of a Hebrew prophet before a Hebrew audience with a Hebraic mindset as proof that he is the Hebrew Mashiach, and then fulfilled that sign through a Grecian reckoning... is utterly absurd!*" Do you trust Scripture more than church tradition?

**Yeshua was resurrected at dusk on the Sabbath day** after exactly three days and three nights entombed in the earth. Any earlier, or any later, would brand him as a false prophet. Can we be certain of this? Yes, the chief priests went to Pilate after the crucifixion, worried his followers would take his body and claim he had arisen (Matthew 27:62-66). "*Command therefore that the sepulcher be made sure until the third day...*" Pilate agreed to the sealing of the tomb until the end of the third day. "*So they went, and made the sepulcher sure, sealing the stone, and setting a watch.*" A cord would be placed across the stone and secured with wax on both ends impressed with the imperial seal. The penalty for breaking such a seal would be death. No one could enter the tomb before the three nights and three days were finished and the guards removed the seal.

There are still a couple of '*Yes, buts...*' to resolve. It says in Mark 16:9, "*Now when Jesus was arisen early the first day of the week he appeared to Mary Magdalene.*" Neither ancient Greek nor Hebrew used commas. They have been placed in our modern translations to clarify. In this case they only confuse. Look at these two versions:

Now when Jesus was arisen early the first day of the week, …

**Now when Jesus was arisen...**
**...early the first day of the week he appeared to Mary...**

*The Hebrew Gospels* say, "*When Yeshua came back to life, ...on the morning of the first day* [Sunday] *he appeared first to Mary Magdalene*" (Mk 16:9). Did Yeshua appear to Mary early the first day of the week, yes he did. "*After that he appeared unto two of them [his disciples] as they walked... Afterward he appeared unto the eleven as they sat at meat...*" (Mark 16:9,12,14). **The context is clear. Mark is listing those to whom Yeshua appeared the first day of the week after he had arisen - not saying the resurrection happened on that day - which would be a contradiction of the prophecy of the sign of Yonah.**

Yeshua had already arisen at the end of the Saturday Sabbath at nightfall - the evening before the dawn on Sunday morning - when it was discovered he was gone. In the HGC, Matthew 28:1 specifies "*On the beginning of day one* [first day] *called 'diumenge'*" (Catalan for Sunday), they arrived and found him already gone. To claim Yeshua arose in the morning light of the first day is a contradiction. The interpretation that does not contradict is the correct one. The Sabbath resurrection [Saturday evening] is supported by both the Greek and Hebrew gospels.

There is still one questionable reference to Sunday as "*the third day.*" The "*third day*" is not contradictory if it includes 3 days and 3 nights, but the resurrection was Saturday evening not Sunday morning as this verse makes it appear. In Acts 10:40, Yeshua appears to two of his family on the road to Emmaus. Acts specified one of them was Cleopas - Yeshua's uncle - and the other was either Simeon (his cousin) or James (his brother).

There are still a couple of '*Yes, buts...*' to resolve. It says in Mark 16:9, "*Now after Jesus was arisen early the first day of the week he appeared to Mary Magdalene.*" Neither ancient Greek nor Hebrew used commas. They have been placed in our modern translations to clarify. In this case they only confuse. Look at these two versions:

1. Now after Jesus was arisen early the first day of the week, ...

2. Now after Jesus was arisen, ...early the first day of the week he appeared to Mary...

*The Hebrew Gospels* say, "*When Yeshua came back to life, ...on the morning of the first day* [Sunday] *he appeared first to Mary Magdalene*" (Mk 16:9). Did Yeshua appear to Mary early the first day of the week, yes, he did. "*After that he appeared unto two of them [his disciples] as they walked... Afterward he appeared unto the eleven as they sat at meat...*" (Mark 16:9,12,14). The context is clear. Mark is listing those to whom Yeshua appeared the first day of the week after he had arisen - not saying the resurrection happened on that day - which would be a contradiction of the prophecy of the sign of Yonah.

Yeshua had already arisen at the end of the Saturday Sabbath at nightfall - the evening before the dawn on Sunday morning - when it was discovered he was gone. In the HGC, Matthew 28:1 specifies "*On the beginning of day one* [first day] *called 'diumenge'*" (Catalan for Sunday), they arrived and found him already gone. To claim Yeshua arose in the morning light of the first day is a contradiction. The interpretation that does not contradict is the correct one. The Sabbath resurrection [Saturday evening] is supported by both the Greek and Hebrew gospels.

There is still one questionable reference to Sunday as "*the third day.*" The "*third day*" is not contradictory if it includes 3 days and 3 nights, but the resurrection was Saturday evening not Sunday morning as this verse makes it appear. In Acts 10:40, Yeshua appears to two of his family on the road to Emmaus. Acts specified one of them was Cleopas - Yeshua's uncle - and the other was either Simeon (his cousin) or James (his brother).

All four Gospel writers constantly outlined the time frame of their narratives. They would cite the season, the day of the week, or phrases such as "*two days later they came to Capernaum.*" The problem is that events were not always related in a strictly chronological order in the Gospels which sometimes skipped about narrating the story. Starting with an anchor point - the feeding of the five thousand - cited in all four Gospels, Rood tied all these chronological references together to nail down all the events of Yeshua's ministry. This had never been done by any previous researcher - a work of several decades!

Rood's analysis revealed the insertion in John 6:4 is completely out of context with the timeline of the other Gospels. It takes place just before the feeding of the 5,000 - which Matthew, Mark, and Luke all confirm took place at the end of the summer, just before the Feast of Trumpets in the Fall. Passover is not in the Fall but in the Spring. The first Passover was recorded in John 2:13. Greek readers, and scribes, were noted for their lack of knowledge of the Hebrew culture and calendar. **Insertions, such as "*a feast of the Jews*," made for Greek readers are common and easy to spot but Greek primacists will not acknowledge them! Should we accept the authority of a later Greek editor of John over Matthew, Mark and Luke?**

If "*Passover was nigh,*" why wasn't Yeshua in Jerusalem for Passover as required for all Jewish believers? Neither in the Gospel attributed to John, nor in any other Gospel, is there a record of Yeshua going up to Jerusalem for the Passover inserted into John 6:4. Indeed, if Passover '*the feast of unleavened bread*' was nigh - the word '*nigh*' means '*near, almost*' upon us - why was Yeshua feeding the 5,000 '*leavened bread?*" For the orthodox in Israel, a week or more before Passover all leavened bread was thoroughly cleaned from households.

The Hebrew word for leavened bread is *'lechem,'* and the word for unleavened bread is *'motzi,'* So in *The Hebrew Gospels* [HGC] we know precisely that Yehovah provided leavened bread, '*lechem*' for this miracle. Not only that but a week after Yeshua feeds the 5,000 - he again miraculously feeds another 4,000 - again with leavened '*lechem*' bread provided by Yehovah!

## Who made a mistake here: Yehovah, Yeshua, or Eusebius?

At this moment in the Gospel narrative, the Pharisees had sent some of their own from Jerusalem to confront Yeshua (Matthew 15:1, Mark 7:1). - **Were these priests skipping Passover as well?** - They were looking for reasons to discredit Yeshua. This would be the perfect '*gotcha*' issue for the Pharisees - the '*Messiah*' feeding leavened bread to his followers nigh unto Passover when all leavened bread was banned! Yeshua would have been finished among the orthodox, reduced to a joke. Yet when the Pharisees confronted Yeshua (Matt 15:2) their only critique was that his disciples did not wash their hands before dinner! The context of John 6:4 could not have been nigh unto Passover.

Eusebius was contracted by Constantine to compile the authoritative version of the Greek Bible from all the competing versions extant at the time. Scribal errors tend to multiply over time as manuscripts were copied and recopied, each time adding new errors and insertions, until the resulting text was replete with mistakes. Most mistakes were minor **but in the case of John 6:4 - the insertion distorted the correct timeline of events in the Gospel record – an irreconcilable textual contradiction!** Nonetheless, it became a doctrinal pillar of the Roman Church.

There are two manuscript traditions extant concerning John 6:4. The first tradition did not include the insertion of '*Passover*' into John 6:4!* In Ms. #472 conserved in Lambeth Palace in London (headquarters of the Archbishop of Canterbury) John 6:3-5 reads, "*And Jesus went up into a mountain, and there he sat with his disciples.* [* no 'Passover was nigh' inserted here] *When Jesus then lifted up his eyes, and saw a great multitude come unto him, he said to Philip, 'Where shall we buy bread that these may eat?'*" Frederich Schrivener, renowned New Testament scholar, is said to have claimed Manuscript #472 was the most important text within the walls of Lambeth Palace.

In the second tradition, '*Passover*' was included in John 6:4 as seen in Ms. P66 & 75. These papyrus texts are said to be from the 2nd century, but research done by Brent Nongbri now places them in the late 4th century. Nongbri's conclusion is crucial to whether John 6:4 was inserted**. The insertion - of '*Passover was nigh*' - into John 6:4 is incomprehensible! In the same breath - as the '*feast of unleaved bread*' is nigh - Yeshua plans to provide '*leavened bread*' to his followers!**

Westcott and Hort, compilers of the modern Greek text of the Bible, postulated that the Greek words, το πάσχα - "*the Passover*" in John 6:4, were possibly "*missing in the Greek texts used by some of the early Church Fathers.* Westcott and Hort, like Schrivener, noticed the insertion of John 6:4 did not match the witness of Matthew, Mark and Luke. **Eusebius either inserted it or canonized a corrupt version -**

**which had that insertion - onto the end of John 6:4 to create his novel theory of the '*prophetic*' 3½ year ministry of Christ!** This is crucial.

There is yet a second doctrinal pillar, another crucial outcome of Eusebius' claim to a '*3½ year ministry.*' Yeshua was born during the reign of Herod the Great (Matthew 2:1 & 2:19). Herod died shortly after Yeshua's birth. **Josephus stated Herod died just after a lunar eclipse, which occurred Jan 10, 1 BC. This is the anchor of the timeline of Yeshua's birth and death!** Yeshua was baptized and began his ministry in his 30th year, as required for any Hebrew Rabbi, to do otherwise would break Torah. He was thirty in the Fall of 29 AD, baptized in the Spring of 30 AD by Yohanon and attended Passover shortly thereafter (John 2:3), his crucifixion occurred on the following Passover in 31 AD. Below is the week of that Passover on April 25, 31 AD.

| Passover 31 AD | Wednesday April 25 | Thursday | Friday | Sabbath | Sunday |
|---|---|---|---|---|---|
| NIGHT | | | | | |
| DAY | | | | | |

Yeshua arose at the end of the Sabbath day, Saturday evening, "*For as Jonah... so shall the Son of Man be three days and three nights in the heart of the earth*" (Matt 12:40). **The 3½ year ministry of Eusebius pushes the crucifixion to 33 AD creating the '*Good Friday - Easter Sunday*' - '*Lord's Day*' doctrine, though he was in the tomb only 1 day and 2 nights!**

| Passover 33 AD* | Wednesday | Thursday | Friday April 3 | Sabbath | Easter Sunday |
|---|---|---|---|---|---|
| NIGHT | | | | | |
| DAY | | | X | | X |

Lucian of Antioch (250-312 AD), compiler of the first complete Greek text of the Bible, "*discerned there were two movements taking shape in Christendom, one loose in doctrine and affiliating itself with heathenism, the other based on the deep foundations of the Christian faith.*" Having established an orthodox doctrine, creed and canon -

Constantine took Christianity center stage as *the* major world religion aided by his star sycophant, Bishop Eusebius. Eusebius' new '*prophetic*' 3½ year ministry of Jesus led to the claim the "*end of prophecy*" had arrived and the conquering-king Messiah had returned in absolute power over all the kingdoms of the earth, in the person of... Constantine!

These two doctrinal pillars; (1) the 3½ year ministry of Jesus and (2) '*Good Friday - Easter Sunday - Lord's Day.*' were now in place. Most churches in the world worshiped God on the Sabbath - Saturday. The new Church, now married to the brutal military power of the Roman State, began to execute Messianic Christians who did not accept their doctrines - and burn their texts of the Bible along with them! Continual crusades were soon launched against these pre-existing Messianic Churches. Many never accepted the authority of Rome to rewrite Scripture - even under penalty of death! These new doctrines were proclaimed as giving Rome justification to use brutal force against Messianics to make them bow to Roman domination. Those who refused were labeled '*judaizers*' and eliminated.

"*When Constantine came to power, Christians were willing to die for their faith - but by the time Constantine died they were willing to murder for it!*"

The Messianic bishops were not invited to the Council of Nicea! Symbolically, Constantine took the bishops of the new Greco-Roman Church up on a high mountain top at the Council of Nicea and promised them they would reign over all the kingdoms of the earth if they would only bow down to him... **and they bowed down before Constantine!**

**They renounced as heretical the original Messianic Church and joined into destroying it! Whether you call the Messiah - Yeshua or Jesus - this was his church!**

**The moment the new Greek Church rejected his Messianic Church - they became apostate!**

# Chapter Twelve

## Columbus & the New World

*If they have persecuted me they will persecute you...*
*because they know not Him who sent me.*
**(John 15:20-21)**

Columbus' final voyage to the New World almost end-**ed his life.** **He was saved by a lunar eclipse** and a fighting force of teenage Messianics. **Columbus was a Messianic!** He went to great lengths to cover his true identity. His voyage was financed, not by Isabella's crown jewels - as has been popularized in legend - but by Messianic Jews who converted to Christianity - and now were fleeing the Spanish Inquisition! The King and Queen had just finished the final *Reconquista* of Spain from the Muslims by taking Granada - their final stronghold - then they immediately turned and expelled all the Jews from Spain. Columbus left Spain at the moment of the final deadline of the Jews to convert or leave Spain, in August 1492. His ships' manifests show a number of Hebrew names. He left with a ship full of Jews fleeing the Expulsion and Messianics fleeing the Inquisition - and a map showing the Indies were West, not East. **This is the untold story of Columbus and his voyage of discovery to the New World!**

Columbus was the son of a Portuguese prince, the Duke of Beja, and a Jewish noblewoman who was a Converso. He was the illegitimate son of the Duke but nonetheless was raised in his household. He was a cousin to King John of Portugal. Columbus gained his superb navigation skills at the school of Prince Henry the Navigator which his father Duke Beja had also attended. It is only by virtue of these advanced skills that Columbus succeeded in finding the Americas and returning alive. It was a far more perilous voyage than most of us are aware.

Columbus changed his name to hide the fact that he was a second-generation Messianic. Even a partial portion of Hebrew blood was enough for the Inquisition to arrest you, convict you, and execute you. The Inquisition preyed upon the wealthy and influential. Not only did he change his name but he disguised his past, claiming to come from Genoa in Italy. Columbus left a colony of 39 people in the New World on his first voyage - many of them Messianics.

**How do we know these things? ...Read on!**

# Columbus was Messianic!

"*Columbus came from the kingdom of Aragon and his native language was Catalan... Christopher Columbus' origins were not obscure by chance, but rather the result of the famous explorer's having purposely hid the fact that he was a 'converso' **- a Jewish convert to Christianity!***" So, he was Catalan - or maybe not - the truth is that Columbus was one of the most secretive personalities of history. Washington Irving, who was paid by the Italian Consulate to write Columbus' official English biography, dutifully reported that he was born in Genoa, Italy. Irving also provided detailed evidence that he was of Portuguese nobility and like his father, the Duke of Beja, had studied at the school of the famous Henry the Navigator in Portugal. **Columbus himself destroyed the records of his family history!** Was he trying to hide he was the bastard son of a Portuguese Prince and a Jewish noblewoman? Or was it that being a Converso of Jewish blood, even second or third generation, could expose him to the Inquisition? On his deathbed, his sons Diego and Hernan begged him to reveal to them their true ancestral roots - but Columbus took his secret to the grave... or so he thought. His sons recited a Hebrew Kaddish prayer at his death.

Columbus could not speak Italian nor did he ever go to Italy in all his travels to the courts of Europe. There is no known relative nor descendent of Columbus from Italy. Columbus could speak and write well in Portuguese - and in Hebrew. One cannot escape certain conclusions springing from these facts. How did Columbus gain entry to rub shoulders with royalty all over Europe? The simple answer is that he was one of them. Illegitimacy did not necessarily prevent a scion of noble blood from assuming the title to which he was due by parentage. In 1978, Patrochilio Riviero and Mascarenas Barrero, two prominent Portuguese researchers, declared Columbus to be the son of Duke Beja, a price of Portugal, which would make Columbus the cousin of the king of Portugal, King John the Second. He was the son of Prince Don Fernando - the Duke of Beja - and Lady Isabel Gonzalves Zarco, a noblewoman and a Jewess, the daughter of the famous Jewish-Portuguese navigator Joao Gonzalves Zarco. **His name was Salvador Fernando Zarco which he changed to Cristobal Colon - aka - Christopher Columbus.**

**Columbus was born in Cuba, Portugal - in the Duchy of Beja.** One of the first of the islands that Columbus encountered - he named Cuba. Not only that but more than forty place names designated by Columbus himself on the island of Cuba - were place names from around Cuba, Portugal. **More than forty place names in Cuba with the exact same names of places in the Duchy of Beja, Portugal!** Keep in mind that these are Portuguese names - not Spanish or Italian names.

When Columbus was first presented to the King and Queen of Spain, he wore the triple pomegranates from the coat of arms of the Duke of Beja. Columbus was raised in the household of the Duke of Beja - as his son - otherwise he would not have been permitted to wear the colors of Beja. A false link to nobility would be easily uncovered and would have ruined Columbus' chances of royal support. It was rumored that Columbus was an agent of the Portuguese king, John the Second, who was his cousin. When Columbus returned from his discovery of the New World, he landed first in Lisbon, Portugal - where he spent several days with the Portuguese King John before returning to Spain.

It was the wife of the Duke of Beja who ended up helping negotiate a treaty between Spain and Portugal that divided up the uncharted territory of South America between them. Strange as it might seem that the Duchess would adopt a bastard son - these things happened more often than one might suspect. So many children died young - having a surviving son could make all the difference between keeping your ancestral title and lands - or having to relinquish them to the crown because you have no heir.

In that treaty of Tordesillas, in 1494, King John the Second of Portugal demanded the boundary of demarcation be moved 370 leagues West, more than 1000 miles. How might the Duchess of Beja and King John have known how to swing such a negotiation? **The very existence of Brazil was not even known at the time... <u>except to Columbus</u>!** This was a master stroke for Portugal, aided and abetted by their native son! Within the Archives of the Indies in Seville, a safe conduct pass was found, signed by King John of Portugal. It was brought to Spain by Columbus when he first arrived. Long thought lost, it declared Columbus to be a friend - and agent - of the crown of Portugal.

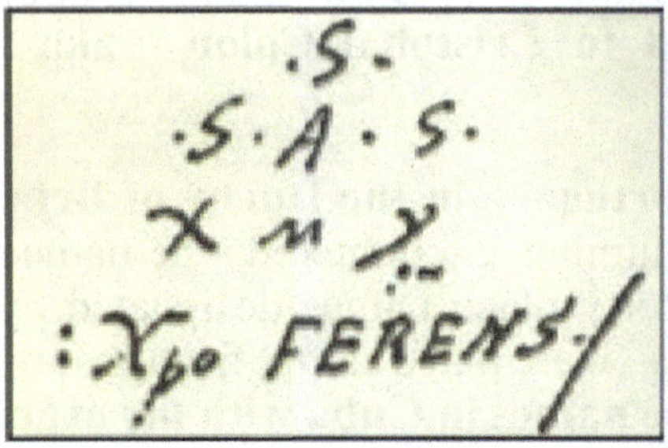


**Columbus signature is a fascinating cipher, or cryptogram. It is in both Greek and Latin - read in two ways:**

1. The Greek letters XRTO are the Templar acronym for Savior - *Salvador*. Feren is short for Fernando, his father's name. The final S is a backward Z for Zarco, so it **reads Salvador Fernando Zarco**, his true name.
2. To read it the other way, note the '*colon*' at the bottom, like in English, followed by Xo...fer, the X for '*Christ*' plus '*ofer*' **- Colon, Christopher**.

The Duke of Beja was a Knight Templar who had gained his seafaring skills under "*The great Prince Henry the Navigator* [who] *had died in 1460. But his crusading spirit lived on, vibrantly, inspiring Portuguese navigators to press on further and further south to the magical passage around Africa to the spice kingdoms of India and Goa. That spirit trumpeted their motto: 'Crusade, Knowledge, Power.' Of these, 'Crusade' was the first and most important commandment! The cult of Prince Henry the Navigator was the seagoing arm of the Templars... Christianizing infidels was the highest Christian calling!"*[5] The red cross so prominent on the caravels of Prince Henry, and Columbus, was the Templar Cross. Columbus had been sent to the school of Prince Henry to get his credentials as a Navigator in the tradition of his father Count Beja, and grandfather Joao Gonzalves Zarco.

Of course there were other motivations for these mariners, **"*The search for the gold of Black Africa and the muscle of black slaves was always subsumed in the higher and more noble -sounding goal of***

***Christianizing the infidel.*"** The skill and daring-do of these mariners and the commerce which followed them had made Portugal the most affluent country in Europe - and Lisbon the busiest port. Now, finding a sea route to the spices of India, avoiding the lengthy and expensive camel caravans crossing the desert sands to the Mediterranean, was their Holy Grail.

"*The lure of spices was more for their use as a food preserver than for their lively taste. To find a way to obtain spices for a reasonable price became a national goal.*" Think of salt - one of those spices in short supply - once it became readily available - the population of Europe tripled! Producers could preserve meats and other foods giving time to get them to market. But for Columbus, his correspondence reveals another motivation, more important than spice, slaves, treasure, or the Roman Church.

## Messianic Zeal

Simon Wiesenthal, the famous Nazi hunter, wrote a book called *Sails of Hope,* in which he researched and made the case that **Columbus was motivated by a deep Messianic zeal to find safe refuge for the many thousands of Jews fleeing the Expulsion,** and the many thousands of Messianics fleeing the Inquisition. One very real reason to believe this evidence is that it was - by far - the overriding concern of practically all Jews and Messianics at that moment in time. It certainly was the principal motive for Columbus' supporters who financed his voyage. Columbus gained an audience with King John of Portugal who was galvanized by Columbus' proposed voyage! But his junta of navigators did not want to change their focus from reaching the tip of south Africa which promised a sea route to the Orient.

King John and Columbus both knew the highly speculative and financially risky voyage could be better financed in Spain. Spain was expelling all their Jews - and their Conversos were fleeing the Inquisition in frightening numbers. Neither were allowed to take their money or possessions with them. **There was a mountain of money - in Spain - to finance a voyage to find a safe haven for the Jews and Messianics!** King John gave Columbus a safe-conduct pass - basically a letter of introduction to the Spanish court - where Columbus "*quickly established himself among the most prominent Converso figures at court.*"[9] As planned, "***The voyage of Columbus was** not funded by the royal house of Ferdinand and Isabella. It was rather **financed by***

***prominent business men who were Messianic Jews... Their names were Louis Santangel and Gabriel Sanchez."***

## The Financing of Columbus' Voyage

Ferdinand and Isabella were indeed intrigued with Columbus' proposal to reach the Orient by going west rather than east. Queen Isabella had granted Columbus a retainer and asked him to remain patient. They were in the midst of a war to expel the Muslims from Grenada, their last stronghold in Iberia. When that was over they could talk about Columbus' plan.

Granada fell January 2, 1492. Within two weeks Columbus was summoned before the Crown. He had been given a stipend for new clothes and a mule to ride. Columbus had honed his presentation to a fine polish but after giving it to the two sovereigns there was an uncomfortable silence. Finally, Ferdinand spoke. The war against Granada had depleted Spain's treasury to the point that no expensive enterprise of this nature could be contemplated at that time. Columbus left them angry and disgusted. He spoke to his Converso supporter, Luis Santangel - the royal treasurer - outside in the corridor and informed him he would be leaving for France, and then to England, to procure the funds he needed for his voyage.

Santangel immediately gained an audience with the Queen. He was eloquent in reinforcing the great promise of the proposed voyage and then, alluding to Columbus' determination to seek funds from Spain's opponents, he concluded, "*It would be a great damage to Her Crown and a grave reproach to Her Highness if any other prince should undertake what Columbus offered Her Highness.*"

## The final point was Santangel's offer to obtain the financing himself. Columbus' voyage was approved!

Columbus had demanded hereditary rights to the lands he discovered and claimed for Spain. Within weeks, the Crown had issued the expulsion order for all Jews to convert to Christianity or leave Spain. The Jews had four months before the deadline. Columbus "*lived in a largely Jewish and New Christian world of navigators, cartographers, astronomers, and mathematicians... Iberian Jews and conversos assisted Columbus in developing his Enterprise of the Indies.*" The Conversos trusted

Columbus - not the Crown - financing would not be forthcoming if his rights were not granted. King Ferdinand blinked first, giving Columbus hereditary rights to any lands discovered.

Was safe haven for Jews and Messianics Columbus' primary motivation? That depends upon whether he was a Converso - or not! ***"Columbus sometimes dated his correspondence by Jewish reckoning from the date of the destruction of the Second Temple… rather than by the birth of Christ….*** [not to mention] *the strange cipher used in all of his letters to his son except the only one Diego was told to show to the Queen... the Hebrew abbreviation for 'beEzrat haShem' ('with the help of God') that is commonly found in Jewish correspondence*." There are various other decisive indicators, "*His staunchest supporters... were Jewish - he seemed to prefer the company of conversos…* [though he had] *to be careful of scurrilous rumor and ambitious Inquisitors. Most definitely!*"[14] "*It is clear that he had an unusually strong interest in Jewish matters, which would be understandable and indeed common among conversos but definitely atypical for a Spanish Old Christian*." Although the financing of Santangel and Sanchez was critical - if Columbus was caught - he ran the risk of burning at the stake for judaizing!

Six identifiable conversos were aboard the three vessels that left for the New World on the night of the expulsion of the Jews. Others might have been present as well, but it was typical for conversos to hide as much of their Jewish background as possible in order to evade the scrutiny of the Inquisition… Perhaps Santangel [& Sanchez, whose son was on the voyage]… shared the dream of many contemporaries that Columbus might find a kingdom of the Lost Tribes, a possible asylum for Jews far from the reaches of the Inquisition. [brackets mine]

Columbus married well, "*In 1478 he made a decent marriage, Filipa Perestrello e Moniz may have been no beauty, but she belonged to a prominent family in the second rank of Portuguese society... close to the royal families of Portugal... Filipa's maternal grandfather had fought alongside Prince Henry in the great Portuguese victory at Ceuta... in 1415. Her father, Bartholomew Perestrello, had been an important explorer… Italian blood ran thick. Her father's family had roots* [in]… *Genoa,*"

**His wife's family history provided Columbus with a perfect cover for his Genoa '*origin!*'**

**In Columbus' journal, he makes it clear where his motivations lie.** "*I could feel His hand upon me. It was the Lord who put into my mind the fact that I could sail from here to the Indies... No one should fear to undertake any task in the name of our Savior, if it is just and if the intention is purely for His holy service.*" Columbus wrote a book entitled the *Book of Prophecies*, which examined in great detail the biblical prophecies concerning far distant lands, undiscovered tribes, migration of peoples - focusing on the Covenant given to Abraham, Moses and Joshua about the Promised Land. Yehovah had promised to bless them - as a nation - if they would enter the Promised Land and make of it a nation of God.

The earliest English colonists, "*The Pilgrims and Puritans actually referred to themselves as God's New Israel. But it was not that they thought they, and the Christian Church, had 'replaced' Israel... They saw themselves as being called into a direct continuation of the covenant relationship between God and Abraham.*" As Yehovah had said to Abraham, "*Go... into the land that I will show you. And I will make of you a great nation, and I will bless you*" (Gen 12:1-2). "*And I will establish my covenant between me and you and your descendants after you*" (Gen 17:7). From the evidence, it seems that this was in Columbus' mind from the beginning. In any case, historians claim that "*American government owes its inception to the covenants of the first churches on her shores.*"

> On the morning of August 3, 1492… the Santa Maria had cast off her lines, set sail, and was gliding down the river with the ebbing tide. As Columbus' ships reached the place… just before emptying into the ocean, a last shipload of Jews was also waiting for the tide. They too were leaving now, bound for the Mediterranean and the lands of Islam... None of that shipload of forlorn exiles could have dreamed that the other three ships on the river were leading the way to a land that would one day provide a welcome haven to their people.

## The Voyage to the New World

There are only three ways in which sailing ships can move, other than being towed by rowers - always a last resort. The Captain must be a virtuoso of the winds, the tides, and the currents! He must know the harbors and the reefs and safe passages of any coastline that he sails. Now a new mastery has been added. He must be able to navigate across the vast stretches of ocean without any landmarks. The sun, moon, and stars, his compass, and astrolabe are his only guides. Columbus was just such a virtuoso! This is potent evidence that Columbus did indeed study at Henry's school of navigation. He was on the cutting edge of seamanship! These were uncommon skills and Columbus would need every bit of his formidable talents to accomplish his mission, and to survive it!

Prince Henry had dramatically advanced the science of navigation. One could approximate their distance from the equator, that is their latitude, by the distance of the sun and stars above the horizon at certain times of the day and night. Now, one could compute their approximate longitude by sighting the North star through view ports in the astrolabe. It was not a precision instrument yet - but it helped! The astrolabe had made its way to Europe from Muslim Spain. By the 12th century there were a handful of expert treatises on the astrolabe in use. "*The oldest surviving, moderately sophisticated scientific work in the English language is a Treatise On The Astrolabe, written by the English poet and philosopher Geoffrey Chaucer* (1343-1400)."

Columbus sailed southwest by south to the Canary Islands, seemingly out of the way. He had a plan that demonstrated his advanced knowledge of the seas. The prevailing winds in the northern ocean were called the *Westerlies*, from the west, but as one went south the prevailing winds were coming from the northeast, called the *Trade* winds. From the Canaries he could catch the Trade winds to take him West then sail north to catch the Westerlies for the return voyage. It was genius! No one had ever thought of this before - for a good reason. **Neither Columbus, nor any man of his crew, had ever been more than 300 miles from the mainland! They sailed from the Canaries on Sept 8, 1492, and a month later, by virtue of the powerful Trade winds, they were more than 2000 miles from Europe!**

## Mutiny Onboard

October 9, 1492, Columbus' two other captains, the Pinzon brothers, Martin and Vicente, demanded an emergency meeting. Columbus was not pleased. It would require all ships to stop by steering into the wind so the captains could row over to the Santa Maria. **The captains were delivering an ultimatum! If Columbus did not turn back soon the sailors would mutiny**, take over the ships, and turn back themselves.

Columbus was faced with giving up his dream. He was angry and frustrated, but he too was aware of the sullen defiance of the crew as the Trade winds pushed them faster and faster, further and further, from their homes and the hope of ever getting back alive. Their minds were full of sailors' nightmares of the sea. What was out there? Was there an edge that one went over like a waterfall with no end? Were there monsters further out in the ocean than anyone had ever been? They knew of whales that could easily overturn their small caravels - what other frightening dangers awaited?

Columbus negotiated three days more to sight land or he would turn back. He felt he had no choice. The danger of mutiny was too real to ignore. He offered a 10,000 maravedis reward for the first man to spy land, almost $7000! The next morning Columbus recorded in his journal that they had made an unbelievable 59 leagues (about 180 miles) during the past 24 hours! The crew grew ever more alarmed by the speed of the ships widening the chasm between them and the safety of land! The nerves of the men were at a snapping point. That day debris from land was sighted, a branch with blossoms.

The tension broke! The men jockeyed to take turns aloft as lookouts. That night they sighted a tiny light far in the distance! Not all the men were convinced that land was near. On the third and final day, at 2 am, the excited call was heard coming from the Pinta, "*Tierra!*" ('*Land!*') In the moonlight one could see a white cliff ahead on the horizon. Columbus slowed to avoid the ever-present reefs near shore. The hours dragged as they approached the unknown coastline cautiously. They reached the island just as the sunrise broke over the horizon.

***"A new day was dawning, and with it, a new era for humankind!"***

# That Thy Holy Name May be Proclaimed

Columbus was the first to set foot on the New World, on the island which he named San Salvador for the Holy Savior. They planted an eight-foot oak cross on the shore. Columbus prayed, ending with this sentiment: "*You have created the heaven and the earth, and the sea; blessed and glorified be Thy Name, and praised be Thy Majesty, which has deigned to use us, Thy humble servants, that Thy Holy Name may be proclaimed in this second part of the earth!*" **Columbus was a Messianic who knew Hebrew and the Scriptures. He knew the Sacred Name of Yehovah, and one must imagine he did what he said, declaring the Sacred Hebrew Name of Yehovah on the New World as he took those first steps!** In a letter to Queen Isabella, Columbus declared, "*I believe that earthly paradise lies here which no one can enter except by God's leave.*"

On San Salvador, Columbus encountered the native Indians who were eager to trade or simply give of what they had in a gesture of friendship. Columbus described them as beautiful and friendly people in a paradise of exotic fruits, plants and animals.

> They ate food that no white man had ever tasted before - sweet juice-giving fruit, & corn, and a pulpy bread made from cassavas. Most of all… they could now drink pure spring water.
>
> The scents of the rain forest were intoxicating, and large birds with plumage of bright reds & yellows & greens filled the air.

It seemed so much like paradise that Columbus himself said he never wanted to leave. **Ah, but what is paradise without gold, silver, and precious stones? These are the things that make sovereigns swoon!** The discovery of gold was quickly realized - to the everlasting sorrow of the inhabitants. To please these new arrivals the natives led them onward in their quest. Columbus sailed completely around the island of Cuba only 90 miles from the mainland of Florida. The natives led him to Hispaniola, now the island of Haiti and the Dominican Republic - where natives wore ample gold jewelry and masks. Columbus traded with them in exchange for a chest full of gold.

When it came time to return, 39 crew members volunteered to occupy the first colony in the New World, there on Hispaniola. The Santa Maria had run aground and had to be abandoned. Columbus returned on the Nina.

Upon the return, the ships encountered a storm worse than any they had ever experienced. The storm separated the ships. It blew viciously for a week, taking the Nina and dropping her in the Azores. The first crewmen ashore were arrested by Portuguese authorities. Columbus managed to secure their release. Getting under sail again the Nina endured yet another week-long storm as violent as the first. This second tempest drove them for another week ripping their sails to shreds. Driven by the immense storm the Nina found itself tossed onto the Portuguese coast, just north of the Lisbon River. It was here Columbus had his finest hour - displaying the most incredible courage and skill!

> The little vessel with no canvas... was blown ever closer to its doom. The storm was peaking in intensity. In minutes they would be dashed to pieces on the rocky coast. With huge waves breaking on the shore there was no way any of them would survive. One slim chance remained... if they could make it to the river's mouth!
>
> But that would mean they would have to take the wind almost broadside - a dangerous maneuver even under the best of conditions. To attempt it in this monstrous sea without sails was suicide! The Admiral gauged their drift and ordered the helm over accordingly... Carefully noting their speed through the water and the action of the waves at the river's mouth, he called out constant corrections to the helmsman at the tiller as he compensated for the ship's yaw. He had to shout to make himself heard above the din of the storm - he was at once calm and exhilarated - challenged at the limit of his God-given abilities...
>
> **It would all be decided in the next minute!**
>
> ***"Lean her to starboard! More! That's it - hold her there - now steady, steady as she goes. Now! Hard a-larboard! Hold her, hold her!"*** **The whole ship groaned and heeled over so far that the sea began to come over her gunwales. The men screamed! It looked as if she were about to go all the way over. But she held and then slowly straightened as a giant wave lifted her and hurled her into the river's mouth!**

After a three-day visit with King John of Portugal, Columbus had restored his canvas and set sail for Spain. There he was showered with money and honors beyond anything one might imagine. He entered the court and told his narrative of the paradise across the seas... then he brought his Indians in with their feathers, paint, piercings and spears - brightly colored parrots on their shoulders. **At the climax, Columbus opened the chest of gold before them. Ferdinand and Isabella fell to their knees before Columbus, and the court, and thanked God!**

It must have seemed God had rewarded them amply for their bloody crusade against the Moors and the brutal Inquisition against the Messianics. And so it has been narrated in church history - but the truth is a more complicated story. Columbus returned on his second voyage the following year to find his colony wiped out. The Indians he had taken to Spain, who could now translate the natives' reports, told the story:

> No sooner had the Nina departed the year before, than the Spaniards who were left behind had started indulging their lust with Indian women. Nor were they satisfied with one each but took as many as they could get. No longer did they barter for gold. They simply seized it, doing violence to any Indian who protested. Quarrelling among themselves and killing one another, they [the crew] had split into factions, and were thus easily ambushed and overrun.

There were factions among the colonists, about 15-20% were Messianics. Some of the crew tried to stop the killing, restore order and decency. It was a harbinger of things to come!

This is the true story of the Spanish conquest - although the Portuguese would show themselves to be even worse. Where did this behavior - of rape, robbery and murder - come from? Were these not Christian soldiers dedicated to bringing the Savior to the infidels? In all truth, they were a perfect reflection of the Roman Church! The blessings of salvation were now being dispensed by the sword! Was this not exactly what was happening in every city of Spain, and soon to be in Portugal**? Fellow Jewish Christian believers were tortured, women and possessions taken at will. After all, they were heretics, and like the infidels, their lives were forfeit - all protest met with slaughter!**

**Upon Columbus' return from his third voyage he was arrested!** The crown had figured out that he had served as an agent of King John of Portugal. Rather than treason, he was accused of mismanaging the island domains of which he was Viceroy. During this time Columbus put together *The Book of Prophecies* with the help of his religious counselor, the very learned monk Gaspar Gorricio. *The Book of Prophecies* was "*a well-constructed piece of propaganda designed to defend himself against the charges made against him and, above all, to justify his own actions... It's purpose was to locate within the historical schema of the salvation of the human race, the discovery of the Indies, presented as the first step toward the liberation of Jerusalem and the Holy Land from Muslim domination, and to assign a prominent role in these events to Christopher Columbus.*" It also attempted to prove, in the process, "*that* *universal domination must be granted to the crown of Spain.*"

This "*form of messianic political propaganda in favor of the crown*" was characteristic of the times. Columbus was released even before *The Book of Prophecies* could be published. He was too big a jewel in the crown of the monarchs to tarnish, much less destroy, but his financing was cut. By his fourth - and last voyage - **Columbus was stranded in Jamaica - his thirdrate ships no longer seaworthy enough for the return!**

Columbus had sailed to Panama and bartered with the Indians there for several dozen gold medallions. His younger brother Bartholomew and his son Fernando were with him on this voyage. He made it as far as Santa Gloria Bay in Jamaica before he had to beach his two ships. They were springing so many leaks they would not make the journey across the Atlantic. "*His worm-eaten caravels, described by his son as, 'more full of holes than a bee's honeycomb,' barely made Jamaica.*"

Columbus was at the lowest point of his entire life. His older seamen mutinied, led by the captain of the second ship, Francisco Poras. Poras rallied the men to mutiny and escape, convinced that Columbus had stranded them in Jamaica rather than going on to the colony at Hispaniola. Columbus let them go.

> The rebel-leader seized the dozen canoes Columbus had bartered from the Indians. Forcing the natives to row, the rebels made three attempts to overcome the fierce currents of the 108-mile-wide channel to Hispaniola. On their final try, they gave up, though only after throwing eighteen Indian paddlers overboard, and chopping off the hands of those who clung to the side. Five months later, after a two-week march across the island, marked by rape and pillage, they were encamped in an Indian village a half mile from Santa Gloria, intending to seize the admiral's ships...

Columbus was in complete despair when he learned the mutineers were preparing for battle against him. His younger brother Bartholomew convinced him to let him lead a counter attack. Columbus had about fifty teenage loyalists remaining.

**They were sons of the Messianic families who financed the voyage.** Not only did they have a fiscal interest in the outcome, but "*their families, being wealthy conversos, were targeted by the Inquisition... to keep their sons safe, Columbus' backers persuaded him to take them* *along*." They surely knew, or suspected the secret, that Columbus was one of them - a Converso!

Bartholomew armed the teenagers and they advanced on the older seamen. The mutineers laughed at the approach of the teenage Conversos. However, their arrogance was misplaced. Bartholomew and his teenage Conversos slew the first six rebels who attacked them and Bartholomew soon had Poras at sword point. The other mutineers surrendered and were disarmed. Poras was put in irons in the hold of the ship.

Columbus' next problem was the natives. After their brutal treatment by Poras' mutineers, they refused to provide any more food to Columbus and his crew. **Columbus had Zacuto's Almanac which calculated a full lunar eclipse was due February 20, 1504. Columbus bet on the accuracy of Zacuto and set up a 'feast and palaver' for the Indian chief Huero and his tribe on the evening of the lunar eclipse. "*Attend tonight the rising of the moon. She will rise inflamed with wrath, signifying the punishment God will visit upon you.*"**

"***When the eclipse began shortly after sunset, Columbus retreated to his cabin.*** *'The Indians grew so frightened,' wrote Fernando, 'that with great howling and lamentation they came running in all directions to the ships, laden with provisions, and praying for the Admiral to intercede with his God that He might not vent His wrath upon them.' When the moon was in full shadow, Columbus emerged. He had pleaded with his God, he told them, who agreed to forgive them as long as the Indians kept the Christians supplied. As proof, 'they would soon see the moon's anger and inflammation pass away.'*"

When Columbus was at his lowest point, stranded on Jamaica for almost a year, he wrote a letter to Queen Isabella that is absolutely chilling in the power of its prophetic voice:

> We have been confined 10 months, lodged on the decks of our ships. My men have mutinied. My brother, my son, and those that are faithful are sick, starving and dying. Governor Ovando of Santo Domingo has sent to see if I am **dead** rather than to carry me back alive. I conclude your officers intend my life should terminate here.
>
> **Should I die in Jamaica and my proprietary rights with drawn, <u>ingratitude will bring down the wrath of Heaven, so that the wealth that I have discovered shall be the means of stirring up all</u>**

**mankind to revenge, and the Spanish nation shall suffer hereafter!**

It was from that very shore Columbus was standing upon that the '*wrath of Heaven*' would indeed be brought down upon the Spanish nation! **Jamaica would become the base of the fleets of the dreaded Buccaneers, financed and led by Messianic privateers, merchants and ship owners - that would bring the Spanish Empire to its knees!**

No doubt Columbus believed he would meet his end on that island. It is hard to imagine him saying such bitter but honest words to the Queen unless he thought they were his last. It was indeed for the best that Isabella never received the letter since he had no way to send it. The rescue ships did arrive and Columbus returned to Spain where he died not long afterwards, in 1506.

Only Columbus' legend remains - and the truth - as we can piece it together. **Columbus' survival due to his teenage Converso fighting force and the lunar eclipse is only one of the many untold stories in the Messianic saga of the New World.** Hopefully, this research will crack open the door to truth and more of those stories will be told.

Even with the sketchy information we have, it is clear the Messianics were not the brutalizing force that the Old Christian Spaniards were. Columbus always traded for gold. He never seized it, nor raped and pillaged the natives. Columbus' attitude is spelled out in the archives obtained by Washington Irving:

**I would not rob nor outrage the country since reason requires that it should be settled, and then the gold may be procured without violence.**

The greed for gold, and the brutal oppression of native peoples, such as the burgeoning and lucrative African slave trade, were hidden beneath the pious facade of "*christianizing the infidels...*" "*It was seen as an act of charity, guided by the will of God, even as the Africans violently resisted the brutality of their purported saviors.*" The Church profited from it, as they did from the brutal oppression of their own Messianic congregants. The Spanish Inquisition followed closely upon the discovery and conquest of the Americas. The Inquisition was worldwide and continued to pursue, imprison and execute Messianics from Mexico to India.

**Columbus, and his heirs, over the course of the next century, forbade the Inquisition in any lands under their authority!**

# Chapter 13
## Marriage at Sinai
### by Catherine Stewart-Jones

Land of Midian, Saudi Arabia
March 14, 2023

I became a believer in the significance of celestial events in the heavens on March 14, 2023 when I was married at the base of Mount Sinai in Midian. **The miracles - celestial events - that were unleashed after the Ketubah ceremony will be forever fixed in my memory!** The Ketubah was a marriage covenant between Yehovah and His chosen people which is outlined in chapter 19 of Exodus and the succeeding chapters. This was the first time in 3500 years that the Ketubah ceremony had been re-enacted - and the only time *ever* done there by a New Covenant group of believers. However, the story did not start there, and it is a story that does not end there. Of that, I am certain!

As it's been said, time is thought to be cyclical in some Eastern cultures, while in the West we see it more as linear. But could it be both? We often see events – even fulfillment of prophecy – being repeated throughout history on different levels, as if using the same template. The bottom line is that our Creator has a master plan, and history is unfolding

His great redemptive plan. We are not to fear the future because Yehovah, the One who made time, has given us the Scriptures describing the beginning, the middle and the end. This helps us understand the times and seasons we are in. Yehovah has already prepared the footsteps of our lives. Every single one of them is known to Him, and His timing is precise and calculated.

For two decades, Dr. Jones had been researching the earliest alphabets on stone, tracing their origin to Arabia and the Exodus. His book, *Writing of God: The Secret of the Real Mount Sinai*, documents these findings. In October of 2022 he was asked to partner with Living Passages, and lead a tour to the Mountain of God, called Jabal al-Lawz (Mount Sinai), and the surrounding areas in Midian. The tour was scheduled for March of 2023. Tourism in the Kingdom of Saudi Arabia was, and still is, in its infancy - but with the ever-increasing political stability of the region – the northwest corner of Arabia has begun to divulge its buried secrets.

Dr. Jones asked me if I would like to go on this tour and research expedition. It was not a hard decision for me. OF COURSE I wanted to go to the real Mount Sinai! This journey would give the opportunity for Dr. Jones and his team, including me, to study the land, inspect the evidence and document it. We would be in the genuine locations as depicted in ancient manuscripts and see antiquities that had remained behind the veil. It would be possible - if we were able to get enough people to join us so that our costs were met. Or, it would be possible simply by the perfect will of our Father in whatever way He chose to make it possible. So I made this request to Yehovah that I would be one to go on this journey to His holy mountain - Har Yehovah.

At this time, Dr. Jones and I were engaged, but I was reluctant to marry. My previous marriage had been abusive and had caused a great deal of pain and trauma. This experience had the final punctuation of seven months in the Kerrville Women's Shelter in Texas. Once I became independent in my living again, I had told close friends to slap me if I even showed the slightest sign of interest in any man again!

People always ask, how we met, so I'll tell our story here. I met Dr. Jones in Kerrville one summer evening in March 2020 while at a friend's birthday party. This was just days before Dr. Jones had been scheduled to go to New Zealand and Australia on a book tour. It was also at the time of the Covid "plan-demic", and governments of both countries had just closed their borders to non-citizens. So only days before Dr. Jones was to leave for New Zealand, he happened to meet me, a New Zealand woman in the beautiful garden of a local restaurant where I was walking my dog.

On my part, I was not seeking to exchange anything more than a courteous greeting, but Dr. Jones clearly wanted to talk. He had deciphered my non-Texan accent as either New Zealand or Australian, and had thought — *"Wow... if I can't go to New Zealand, then Yehovah has brought New Zealand to me!"*

As I attempted to politely end the conversation, Dr. Jones told me he was an author and had been about to go to New Zealand and Australia on a book tour. A likely story I thought! But what a great way to impress! I felt obliged to ask, "So w*hat do you write about?*"

"*Religious topics*", was his response. Hmmm. I could have stopped there, but I could see he wanted me to ask him "what religious topics?"

"*What kind of religious topics?*" I asked.

"The real *Mount Sinai, the Writing of God, education...*" He didn't have a business card to give me, but he had told me hs name was Dr. Miles Jones. Okay, so I could look him up online and see if he was for real or not.

He continued by asking what brought me to Texas? I answered nonspecifically, "*I go where God leads me*". He asked me what kind of work I did. I told him I was a seamstress. I felt somewhat undone having this man ask me more questions than I felt comfortable to answer. He asked if I had a business card. I gave him one and said goodnight.

Once home, I looked for him online and found he was indeed the Dr. Miles Jones of his claims. I was even intrigued about what he apparently wrote and taught about. I didn't expect he would call me the next day and tell me he had a jacket that needed repairing. Well, maybe I did! He asked if he could buy me lunch at the same restaurant where we had met, so he could bring me the jacket. Smooth move! I found myself saying yes to Dr, Jones, and so we met once more in the beautiful garden of the restaurant.

Sitting at a table under a sun umbrella eating a delicious salad, he spoke about the earliest writing being the "writing of God" in Exodus 32:16, and of where he had discovered T*he Hebrew Gospels* that were not supposed to exist. He also told me that Yeshua was Jesus' actual name and that God's real name was Yehovah. He told me that while Jesus and Yeshua were the same historical person, they had different archetypes. He told me things that I had not known - and I had questions for everything he was telling me.

I first came to faith in Jesus as an eight-year-old. Through the years I had grown in my faith and understanding of God and Jesus. I had even been a Salvation Army officer (minister) for fifteen years, but I had never heard anything like what I was now hearing from this man! It wasn't that I didn't believe him. In fact what I was hearing, arrested my attention. It was like I was thirsting for what I was hearing but hadn't known it until I heard it. I just needed him to slow down some, because it was like trying to drink from a fire hose!

I left Dr. Jones and the restaurant with his jacket and a signed copy of the book *The Coming Crisis*. Good choice Dr. Jones. It was the smallest and easiest to read of all his books. I stopped everything that afternoon and night to read it through.

And so I was convinced of two things. First, since I was on a quest for truth in every area of my life, this man had something I needed to know more about. And second, that God, or Yehovah as I could now call Him by name, had gifted me this opportunity to know important new things - so I would have to approach this very matter-of-factly with no romantic involvement.

## Well, that second conviction didn't hold up!

So here we were. We had actually planned to marry eighteen months after we first met, but just weeks before the wedding date, I got a clear red light from Yehovah to **STOP!** This was devastating for Miles but a huge relief to me. We broke up, but a couple of months later I agreed to get back together with him. For a whole year Miles did not bring up the subject of marriage, which I was most grateful for. Then, it seemed like he was asking me every other week! All I knew to say to him was, "*I can't marry you. Not until I get a green light from Yehovah*". This felt safe and secure for me. I was in no rush and knew I was on solid and unquestionable ground. I knew I must always listen, and try not to miss anything of Father's guidance to me.

At the end of 2022, I started learning Hebrew from our Benai Emunah Hebrew teacher Jenny Webster. I had made a preliminary start with Miles by getting familiar with the Hebrew letters, sounds, various words and names, but I was a Hebrew language beginner. One of the first things Jenny taught me was the Aaronic blessing. Miles thought it was crazy to teach a beginner the Aaronic blessing. It was indeed difficult to learn and memorize. I now don't remember if this was Jenny's idea or mine, but I wanted to learn it and was absolutely committed to learning it.

Jenny was patient and persevering! Over and over and over again I would practice the sounds, the words, the phrases, the lines, then eventually the whole blessing. Why was I so desiring and committed to learning this blessing that even Miles hadn't learned. (I even ended up teaching this blessing to him!)

Initially, I didn't know if it was possible that I would get to go on this archaeological survey of Mount Sinai. It was financially costly, requiring funds I did not have. My friend Janice Baca, the Translation Team Leader, felt similarly. We reassured our hearts that Father would make a way to Mount Sinai for those He wanted there. Miles however was simply confident that I would go. I was seeing the miracle of others having their costs met in incredible ways. One day Janice told me that I should just prepare as if I was going because she was certain I would be. So I did. Janice herself had prayed, "*Abba, please make a way for us to go to Mount Sinai.*" Yehovah heard her petition. On November 3rd, Janice and her husband David, were gifted with an all expenses paid trip to Mount Sinai.

Then there was Danielle's story, another friend who had learned of this upcoming expedition to Mount Sinai in the Benai Emunah newsletter. Her heart had immediately longed to go. As a young mother homeschooling their son, going on such a trip would leave no one to care for him while her husband was at work. She put her desire in Father's hands and spoke nothing of it to her husband. Several weeks later, her husband surprised her by saying, "*I saw a newsletter about a trip to Mount Sinai. I think you should go. I will look after Isaac.*" And so Danielle was to come, another miracle!

I had been sending out information of this epic adventure journey to people, and answering many emails of inquiry. Not only would this be a tour, but an extension week was being offered to anyone who would like to join Dr. Jones on an Archaeological Survey of Mount Sinai and the surrounding areas in Midian.

Neither Covid jabs or testing were required - an assurance we all wanted before signing up! And sign up they did. Very soon we had sufficient compensation funds for my travel costs to be met. That was a joyful day for me!

By now, we had 15 women that would be going. I suggested to Janice that we invite them to a Zoom call. Perhaps the women might even like to regularly meet together by Zoom, prior to our epic journey. It would be a great way to build relationships and know each other prior to leaving, I thought. And so we did. The *Women of Sinai* met weekly, for the two months prior. It was here we shared information and ideas on travel, clothing, climbing necessities and accessories, physical training strategies. It was here we spoke our questions, concerns and wonderings. It was here we prayed and shared Scripture together. It was here we shared how The Spirit was speaking to our hearts. It was here we shared the dreams we'd had and the visions we had seen regarding our Mount Sinai journey. And since Dr. Jones had brought to our attention that Psalms 24 actually referred to Mount Sinai, it was here in our women's weekly zoom calls that we realized Father was gathering His daughters for a special purpose on this trip. To lead in worship and purity of heart. Psalms 24 became our focus, our meditation, our standard and our dreaming. Psalms 24 became the central focus in our spiritual preparations.

## The King of Glory

Psalms 24 - A Psalm of David

The earth is YHVH's and all that fills it - the world, and those dwelling on it. For He founded it upon the seas, and established it upon the rivers. Who may go up on the mountain of YHVH? Who may stand in His holy place? One with clean hands and a pure heart, who has not lifted his soul in vain, nor sworn deceitfully. He will receive a blessing from YHVH, and righteousness from God his salvation. Such is the generation seeking Him, seeking Your face, even Jacob! Selah

Lift up your heads, O gates, and be lifted up, you everlasting
portals: that the King of glory may come in.
*"Who is this King of glory?"*
YHVH strong and mighty, YHVH mighty in battle! Lift up
your heads, O gates, and lift them up, you everlasting portals:
that the King of glory may come in.
*"Who is this King of glory?"*
YHVH-Tzva'ot - He is the King of glory! Selah

While some of the women had many dreams and visions, I had just three short dreams. On January 18th, 2023 I had a dream where these words were spoken to me in a clarion voice, "***There is more ahead of you than time would suggest Fiona.***" I am now known by my middle name Catherine, but Fiona is my given name. I did not know what this could mean, and if this pertained to my immediate future or not, but I was told it was important to take notice when spoken to by name. I kept this divine message in my heart, and whenever it came to mind I tried to remember to say, "*Hineini*". *Hineini* is Hebrew - meaning - I am ready Yehovah. Here I am.

On February 14, 2023 I had a second dream. A **selection of different shades of blue linens were put before me and I was to choose one. I then saw/read the word ISAIAH**.

*Linen* is Yehovah's choice fabric. He created it with extremely high frequency, and because this frequency is so much higher than that of our body, it raises our body's energy and promotes health. "*The priest shall then put on his linen clothes, with linen undergarments next to his body...*" (Leviticus 6:10)

*Blue* can represent Yehovah or something dedicated to a holy purpose. It can also symbolize royalty or riches. It is written the most in the book of Exodus (34 times). *Blue* can also mean or symbolize service to God and godly living. *Holiness.* I did not know how Isaiah might relate, as I could not find a scriptural link between *blue, linen* and *Isaiah,* but this verse kept popping into my mind. "*A highway shall be there and a road, and it shall be called the Highway of Holiness. The unclean shall not pass over it, but it shall be for others.*" (Isaiah 35:8)

During Janice's prayer time one morning in the middle of February, she saw a vision of Esther coming before the king. Why was she given this vision? Then she realized – Purim was approaching. Janice wondered if it would be around the time we were going to Mount Sinai. She was shocked to discover that not only would it be taking place at the time of our trip, but we were scheduled to climb Mount Sinai on the third day of Purim. So, we would be presenting ourselves to the King of the Universe on the same day that Esther presented herself to her king!

Janice called me to share the exciting news. I said, "*Janice, what is the Torah portion for that week?*"

> [The Torah, also known as the Five Books of Moses or the Pentateuch, is divided into 54 separate portions and linked to a specific week of the year. From the time of Ezra till today, these weekly portions are read aloud in synagogues and Messianic assemblies on Shabbat (Sabbath), and speak prophetically even into our modern times.]

She responded, "*Catherine, what is the Torah portion for that day?*" Incredibly, the Torah portion during our visit to Mount Sinai was Exodus 33:17-23 – the passage describing Moses hiding in the cleft of the rock to see the glory of Yehovah pass before him!

> *"And he said, "You cannot see my face: for no one can see me, and live... Behold, there is a place by me and you will stand on a rock and when my glory passes by, I will put you in a cleft of the rock, and I will cover you with my hand while I pass by..."*
> (Exodus 33:20-22)

He passed by, Moses saw His glory, granting his request! We wondered if this was a sign that miracles would happen when we, too, came to Mount Sinai?

As Janice shared her excitement with the group, others decided to join us in fasting and prayer even as Esther's people had joined her. We were preparing to go before Yehovah at Mount Sinai. As Psalms 24 said, only those with clean hands and a pure heart could ascend Har Yehovah. And so we prepared with prayer.

Prayer was not our only preparation. And while this tour was being organized by a travel company, there was also a lot of individual and group planning, organizing and preparation for the journey.

We needed suitable cotton/linen clothing, as well as gloves, hats, shoes, head coverings for the women, First Aid and medicinal items, travel toilet paper and hand sanitizer. To climb Mount Sinai required

- boots, pants, hiking sticks, day packs, camelback water bags for hydration, electrolytes, energy snacks. Then there were all things electronic: phones, cameras, chargers, memory cards, adapters, filming and audio equipment, for filming and photographing. And finally, we needed inscription charts, reference material, pens, markers, small plastic bags for collection of findings, and a whole host of miscellaneous items.

We had Zoom meetings to give information and training to those team members staying on for the week of research in and around Mount Sinai. If we didn't already know, we were to learn Google Earth and how to follow a pin or lay a pin during our journey. A pin is a location that someone has identified as being important. We would otherwise have no idea what we were walking over as we walked through the archeological areas. We would be documenting the co-ordinates of previously found inscriptions, as well as surveying the areas where we would look for inscriptions of the Israelites. We wanted everyone to be as involved as much as they wanted to be.

There was organizing and booking of flights to and from Washington DC, as well as required accommodation while still in the USA. Passports and other required travel documentation were obtained. We needed to organize international calling on cell phones, and for some of us women, we learned to play and practiced our kalimbas - small thumb pianos for worship of Yehovah together.

Important too, was to familiarize ourselves with the rules and etiquette we would need to follow in Saudi Arabia. For example, men were not to engage in a conversation with a Saudi woman. Women were to dress modestly with loose-fitted clothing - knees, shoulders or décolleté were not to be exposed. We were not to eat, pass, or receive food with our left hand. We were not to show the bottom of our feet (with shoes or without) as it is considered rude to sit with your leg up and crossed with the bottom of your feet showing. We were also to be aware that Saudi Arabia had no private or public religious freedom for non-Muslims, and this included us.

So, it was a big job to plan, purchase and pack everything Miles and I would both need. Exciting and fun too! We would be there in spring, and temperatures would fluctuate from mid-80's°F during the day to low 30's at night.

The item I most wanted to take was my ram's horn, or *Yovel*. It is after all, what evokes the power and blessing of Yehovah when it is sounded. Yet we were told we could not bring any shofars as they would be confiscated by Saudi authorities on entry into the country, or worse still, jeopardize our team being allowed entry into Saudi Arabia. While disappointed by this restriction, we comforted ourselves knowing there was a *Yovel* hidden somewhere on Mount Sinai for us to find. And find it we would!

As Janice prepared for Mount Sinai, her friend asked that a white handkerchief be placed upon the *Mountain of God* for the healing of her blind eyes, and this idea went viral! Between the group members, we brought about 200 handkerchiefs to touch the altar of Yehovah on Sinai! Names were either written, embroidered, or left blank for all who would receive a handkerchief. Janice spent weeks embroidering names on each handkerchief. For the extra handkerchiefs, she embroidered the name of Yehovah!

As our departure date approached, excitement grew and spread throughout our Benai Emunah Institute volunteers and the Mount Sinai group. Congruently with these preparations, another team at Benai Emunah was formed led by John Reed-Austin. Their focus was to restore the original biblical songs – the Mizmorim. Their first would be Psalms 24. A global choir of volunteers was quickly brought together by Jonathan Felt, who added their voices to the recording of Psalms 24 completed by the Mizmorim team. This was to be played at Mount Sinai.

On February 27, 2023, just three days before Miles and I were to leave, my friend Renae came to visit. Renae is an anointed daughter of The Most High who shakes the spiritual darkness when she prays and worships. Her love for Israel and the Jewish people is great, and she has made dozens of visits to Israel. Knowing of Miles' and my imminent journey to Mount Sinai, she had driven from her nearby town to make an unexpected special delivery. I was looking forward to sharing my excitement of the wonderful Yehovah-happenings related to this trip, and to have her pray with me for all matters pertaining to this very strenuous time of preparation. She knew it was taking a heavy toll on me, in spite of how wonderful it all was.

Time spent with Renae was always a blessing, but today's visit would bring me special encouragement and joy. She told me she had some special things to loan me for this trip. She then proceeded to take two big silk worship flags out of a bag. Both were gorgeous and amazing, but especially the one that had Yeshua's name across it. Then Renae paused momentarily and said in a more intimate, almost sacred tone, "*the other thing I have for you to borrow is something I have never released to anyone else, but God has told me that He wants you to take this to Mount Sinai...*" Renae carefully took something out of the bag as she said, "*This is my Queen Esther belt. There is a special purpose for you with this. I don't know what it is, but you will find out.*"

Wait, what?... did she just say, "*Queen Esther*"*?* As in... the **'Queen Esther of Purim'?**

This was to be the latest piece of *'incredible'* regarding this journey. I'd had no previous conversation with Renae about Janice's Purim dream. And Renae wasn't aware that we would be at Mount Sinai for Purim! I knew this was going to be good!

I had not heard of an Esther belt before. I was most eager to know more. Renae began to tell that it was a holy garment and that until now, she had not loaned it to anyone. She told me that Yehovah Himself released the belt to me. She put the belt on me. It was very beautiful.... A soft wine-colored sash with small gold bells either end. She had wrapped the fabric twice around my waist. The bell ends hung low on the side of my body. With even just a little movement of my body, the bell's delicate sound could be heard. It was very beautiful!

"*It's worn to bring the essence of Yehovah's presence where He desires dominion*", Renae told me. "*It's a tangible visual, that changes the atmosphere through His Presence and His Directives*". She added a caution, that Yehovah demands obedience, and the belt was only to be worn when He directs, and for His purposes.

Of course I was riveted, as Renae told me the incredible story of this beautiful belt Each time she had worn it - highly significant even miraculous experiences had come with it. With Renae, this didn't surprise me at all. Still, I was gob-smacked.

When I eventually got to tell Renae of Janice's Purim dream, and that we were actually going to be camped at our King's Mountain on Purim, her eyes twinkled as she smiled and said, "*Well... what more confirmation do you need!*"

Renae told me her prayers were that I would have the adventure of a lifetime. She told me that worship was key. She told me that Yehovah would guide me and keep me every step of the way. She told me that I didn't have to be concerned about my physical limitations and that all I needed to do was to "show up". Before she left my home, Renae prayed over me and blessed me. I would not see Renae again till we got back from Mount Sinai.

On March 4, 2023 Miles and I flew from San Antonio to Washington D.C. We would be there four days before the rest of our team arrived. This provided opportunity for Miles to research, and for us to meet with the Curator of the *'Museum of The Bible'*. Alice was one of the women traveling to Saudi Arabia with us. She lived in Washington D.C. near the Dulles International Airport, where we would be leaving from. Her kindness to host Miles and I was a huge blessing and help, as we would require two rooms.

The day before our departure, Alice rang me to check that I had my USA Residency Card, knowing that I'd need it to return to the USA. Oh no... I could not believe that I had left it at home in Texas! I rang our tour company operator and explained the situation. Her response was clear. She told me I should not go on the trip but return home instead. She knew from previous experience that without my Residency Card I would not be allowed back into the USA when we returned. She told me nothing was worth that situation happening. I did not like that advice at all - what a huge blow!

At the time, Miles and I were at the Library of Congress, and Miles was in deep discussion with Dr. Yohannes Zeleke, who would also be on our trip. On this matter I was not reluctant to interrupt. The alarm in my voice was more than an acceptable library whisper and we were directed by the librarian to take our conversation out into the corridor. Here, I informed Miles that I didn't have my Residency Card, and that Rhonda had determined I should not pursue this trip. Instead to go home. Miles immediately declared, *"That's not happening. I need you with me. You're going!"*

So I called Janice and David. They lived 40 minutes from my house. I fully expected them to have already departed for D.C. Instead, they were loaded up in their truck and about to leave for the airport. I told Janice the situation and Rhonda's advice to me. Janice didn't miss a beat and was just as adamant as Miles. She declared, "*Well that sure isn't happening! Yehovah has purposed you to be at His mountain, and you WILL be there!*"

They immediately drove to our town of Kerrville where I lived, and to my house where my neighbor happened to be home - and she had my house key. Janice was able to enter my house and locate my Residence Card. Then she and David still got to the airport in San Antonio in time for their flight to DC from where we were all departing to Saudi Arabia. Praise Yehovah!

The precise timing of all this happening was something to consider. If I had called even a few minutes later, they would have already left their rural property and headed out in a completely different direction. They may not have had cell phone service to even get my call! And should I have simply given into the idea that I shouldn't go to Saudi because of me being refused entry back into the USA at the border, this would have meant forfeiting the place where Father wanted me to be. Where He had very special plans for me.

Together with our film crew, some of our Benai Emunah team, and many new friends, we had a total of 35 people embarking on an adventure that would result in miracles at Mount Sinai. ***Miracles we could never have imagined that Yehovah had purposed.***

The 12-hour flight from Washington D.C. to Jeddah, Saudi Arabia, seemed to be an eternity, so it was great to finally land. We then took a connecting flight to the small airport of Tabuk where our Land Cruiser drivers were waiting for us. This was March 6, 2023.

After stopping at a restaurant for a delicious Saudi smorgasbord, we were taken to the brand-new Millennium Hotel. We were all super tired and it was after midnight before we got to bed. With only a few hours of sleep, Miles arose for his 6am meeting with a Saudi man. The meeting would prove to be mind-blowing. As Miles related to me, "*After introductions, the man immediately reached into his bag and pulled out an ancient book with paleo-Hebrew writing embossed on pages of beaten gold! The first thing I saw was a large circular seal on the cover of the gold book with the Name of YHVH written in the ancient alphabet! That was better than a cup of coffee to wake me up!*"

There were many other books written on lead tablets. These were the archives of the Bene Quareza, a Hebrew priestly tribe fleeing Roman persecution from Israel to Mecca in the second century AD.

And so here we were in Saudi Arabia where the foundational narrative of the Exodus, as outlined in the book of Exodus, contained some of the most vital and epoch-making accounts in all the Tanakh (Old Testament). Our objective was to find the archaeological and geological evidence of the Israelites' deliverance from Egypt, Moses' revelation at Mount Sinai, and the nation of Israel's ensuing desert wanderings. Later, we came to learn that these very stories were related in the Gold Book of the Bene Quareza in the ancient alphabet of the Israelites on the first day of our expedition to Sinai!

In the first week of time in Saudi Arabia, we: explored the Land of Hagar, Jethro and Moses, we went to Jethro's Caves and the well of Moses and Zipporah. We examined and hiked up the majestic Split Rock of Horeb at Jabal Maqla. Here, Moses struck the boulder and Yehovah miraculously provided water from within. It was breathtaking to see! We went to the peak of Jabal Maqla, Moses' Altar, where the pillar stones representing the 12 tribes of Israel are found, and to Aaron's Golden

Calf Altar. We explored the cemetery that is dated back to the time of the Exodus, and saw the standing stones of ancient times that mark the grave sites. We drove through the Land of the Amalekites. We reached Elim – the place of 12 wells and 70 palms by an off-road visit through the meandering open valleys most likely tread by the Israelites. Some of us had the incredible experience of walking from Elim through the Wadi al Tayyib to the Red Sea gateway. We also saw the beach directly across from Nuweiba.

This meant some long hours spent in our Land Cruisers each day, but everything to me was captivating in the dark red windswept landscape. And throughout our time in Saudi Arabia, the NEOM project was in full swing. Neom is the ultra-modern city they are building right beside Mount Sinai.

## Going Before The King

On March 8, 2023 we went off-road to the specially prepared base camp at the foot of Mount Sinai, Jabal al-Lawz (meaning the almond mountain in Arabic) for the night. Jabal al-Lawz is hailed as one of the most mysterious mountains (with Maqla its highest peak) in the Middle East. Most people think that Saudi Arabia is a land of vast sand dunes, but this is far from true. With its summit at 8,362 feet above sea level, Jabal al- Lawz is the highest mountain of the Tabuk Province. Jabal al-Lawz is now known to be the real Mount Sinai. This is where Moses received the writing of God and the covenant commandments, meeting Him in a cloud of fire scorching the mountaintop. And here we were! We settled in our tents and ate the evening meal. Spending the night with the tribesmen out under the stars around a crackling fire with Jabal al-Lawz looming overhead was a sight of awe.

The Saudis had set up a karaoke speaker expecting to play Saudi secular music through the evening. However, our group had a different idea. We gathered singers, dancers, Bluetooth speakers, tambourines, thumb pianos and raced outside the camp to worship Yehovah! With surround-sound speakers at the entrance of our tent compound, we worshipped, danced, and sang Mizmor 24, also lifting up our hearts and voices in prayer. Not one person stopped our worship and prayer for the 90 minutes we were there! And what's more, Yehovah graced us with an enormous full moon over the mountain that night that bathed us all in moonlight! And it was here that I had put on the Esther belt. I believed this was ***the*** occasion I was purposed to wear the belt!

The following day, March 9, 2023 was our **first day at Mount Sinai**. It was also the third day of Purim. The same day that Esther presented herself to the king! Today, we were going to present ourselves to The King!

It was our greatly anticipated climbing the mountain day! And it was climbing, not hiking. We were to be physically fit. The climb was advanced. So in the months prior, those intending to make the ascent had been in physical training. The Plateau of the 70 Elders would take around 6-7 hours round-trip, Elijah's Cave 7-8 hours round-trip, and the Blackened Peak 7-8 hours round-trip. It wasn't possible to get to both the peak and Elijah's Cave, so we had to chose whether we climb to Elijah's cave, the Plateau of the Elders, or to the Peak of Mount Sinai. My desire and determination for the past months had been to reach the Plateau of the Elders, and I had done my best to be ready for this.

After breakfast we all gathered for instructions and the pep talk on climbing the mountain. It had been an extremely cold night in our tents, but now the sun and risen and there was not a cloud in the sky. It appeared to be the most perfect day to climb Har Yehovah! Necessary but deflating was the very stern advice given by Rhonda our tour leader, that no one should attempt to climb the mountain if they had any doubt whatsoever about out their capability or wellness. She reminded us that recently even some accomplished climbers had got into trouble on the mountain and had either jeopardized the safety of their co-climbers or it had meant the whole team had to forfeit going further. Someone had even required an air lift off the mountain.

So while I laced up my boots, put on my camel-back pack with ample water, grabbed my climbing gloves, hat and wet neck tie, I was completely uncertain about my ability to ascend Yah's mountain after setting my heart on this for many months! I knew that those who didn't climb would have plenty to explore on flatter ground and small mounds that mysteriously could hide secrets of the past. Miles wasn't climbing, he was filming, so I considered not climbing too.

Our team drove to the base of Mount Sinai so we could then walk the foothills to Moses' Altar. It was from here that we would all separate for our day's adventures. I was still in my indecision, when I realized that those going to the Peak and Plateau of the elders had already left, and the group going to Elijah's cave was just heading off.

I found myself hurriedly saying goodbye to Miles before running to catch up and get in step with Roland and Lilian Atencio and the four young Saudi men who would be our guides up and the rugged mountain to Elijah's cave.

There is only one cave that is very noticeable on the mountain from the ground level. It is seen on the front of the mountain, above the altar of Moses and the 12 pillars. According to 1 Kings chapter 19, the prophet Elijah stayed in a cave on Mount Sinai and interacted with Yehovah there. He was fleeing Ahab and Jezebel in fear of his life.

Elijah had traveled 40 days and 40 nights to Mount Sinai, and once there, he found a cave to sleep in that night. The next day, Yehovah instructed Elijah to "*go out and stand on the mountain in the presence of Adonai, for Adonai is about to pass by.*"

A great wind then caused falling rocks, an earthquake, and a fire. Elijah however, was protected as Adonai was not "*in*" the phenomenon. Once the dangerous weather subsided, Elijah heard a "*gentle whisper*" from Yehovah and "*pulled a cloak over his face and went out and stood at the mouth of the cave.*"

Immediately I knew this climb would be challenging even with two climbing poles. The whole mountain was rocks and loose stones, and it was steep! There was no trail. But… I was ascending the mountain of Har Yehovah! How surreal. I was living Psalm 24! What a privilege, joy and blessing. It had been my heart's desire to do this, and now I was!

We were very pleased and grateful to have the help of our four Saudi guides who we were there to get us safely up and down the mountain. They appeared to know very little English, yet that didn't seem to be a problem. They were friendly and responsive to our communication efforts.

Once the climbers had set off, Dr. Jones finished a video recording at the altar, and some of our women team members placed dozens of white handkerchiefs on and among the rocks of Moses' Altar. They praised Yehovah and prayed over these small tokens – asking for miracles like those that followed Paul. Janice of course, was one of the women.

*And God wrought special miracles by the hands of Paul: So that from his body were brought unto the sick handkerchiefs or aprons, and the diseases departed from them, and evil spirits went out of them.* (Acts 19:11-12)

Actually, they had surround-sound speakers at the altar and worshipped, danced, and sang – loudly and boldly! Not once did they look over our shoulders in fear. In fact, our American tour guide said she looked for the Saudis and realized they were nowhere to be found!

The climb was indeed challenging. It took us 7-8 hours. There were no trails to follow, we were making a steep ascent on loose stones and rocks. Each of our groups did so with the experience of at least several Saudi men as guides, and we relied on them to scout out the best and safest way up and down the mountain.

Halfway up to Elijah's cave, I realized something was seriously wrong! While I was not out of breath from exertion, and my legs were doing just fine - my heart started to go berserk! With the first sense of it, I slowed down. Then very quickly my heart was experiencing a full attack. I was now at the back of our small group, and with some distance between me and others. One of the guides was with me. The crushing pressure on my heart was severe. I immediately stopped. All I could do was pant with short shallow breaths in an attempt to calm the berserkness going on in my heart and relieve the crushing sensation. What was happening? Is this a heart attack!

The guide who was with me didn't appear to speak English. He seemed to assume I was only tired and just needed to catch my breath. Yet the one thing he did know to say in English was, "T*ake your time, take your time*". I could only very quietly say, "*It's my heart*", but he did not understand and repeated with a smile, "*Take your time, take your time*". After a few minutes my heart settled somewhat, so I very slowly started to climb again. But three more big steps up on the rocks that required all my effort, even with the help of climbing poles, and my heart was again in great pain and going beserk. Was this because of the myocarditis I'd had some years back that had left my heart muscle scarred?

I had not even given this heart history a single thought the whole time of preparing for Mount Sinai and the climb. Even while on the mountain and experiencing this distress, I did not put two and two together. It would only be once back in the USA, I would learn that high altitude is dangerous and even fatal, for someone with residual heart scarring from myocarditis. This is because the heart's function is already compromised, and high altitudes further strain the heart.

Lilian, Roland and the other guides were just far enough ahead that I couldn't explain was happening. They also thought I was just tired and needed to rest, so they weren't concerned but focused on their challenge of climbing. I remember the thoughts and conversations with Yehovah I had while this was happening…

***"I think I could die on Your mountain"***

***"What an incredible place to die!"***

***"Yehovah, if I die here that's absolutely fine with me."***

***"If You want me at Elijah's cave,<br>You will need to get me there."***

And so, from halfway up the mountain till I reached the cave with my guide, I contended with my heart going beserk and having to stop every two or three steep step ups. And still none of the group knew what was happening to me, and still my guide repeated over and over "*Take your time. Take your time*".

As I saw the cave getting closer, my mind was focused on praising and thanking Yehovah that He was indeed *giving His angels charge over me to keep me in all my ways* (Psalm 91:11-12). I was getting there.

Once I got to the entrance of the cave my heart was immediately well. Any thoughts of me, were replaced by the sheer thrill of where I was! I was now below the cleft of the rock where Yehovah put Moses and covered him while His glory passed by. Exodus 33: 21-23. And I was now at the entrance of the cave where Elijah had come and received Yehovah's revelation. The cave where Moses and Abraham may also have been! Perhaps Paul too. THEY had all climbed this holy mountain of Yehovah and possibly come here to this cave for refuge… And now I had too. **HOW INCREDIBLE!**

So few people in the world, so few children of Israel throughout all the centuries, had *ever* climbed this holy mountain or been in this cave. Yet multitudes of people had read many times over about Elijah's cave, only to imagine what it would be like! **How was it possible that I was here?**

Giddy with joy and wonder, I slowly moved into the cave. Wow. It was the size of a small bedroom. It was about 15 feet high, 20 feet wide and 20 feet deep. The top of the cave decreased from 15 feet to about 8 feet as I went further into the back of the cave. I imagined myself sleeping here. It would definitely be safe and comfortable for someone to sleep in.

While I might have had my feet on different rocks and stones as I climbed the mountain to get here, now in the cave I was definitely placing my hands and feet where Elijah and Moses… would have stood, sat, slept, touched, prayed, ate, wept,… These great men of faith… how many people I knew who would want to be **RIGHT WHERE I AM NOW**!

Elijah had surely been tired from ascending the mountain to this cave. I too was now tired and needed rest, food and water. Here now in March 9, 2023, I, Fiona Catherine Stewart, was finding refuge in Elijah's cave.

Looking out from the mountain I was able to see another viewpoint of the standing stones of the ancient cemetery – dated to Moses' time. I got out my phone to take photos and to video the cave and out across the great vista. It was awe inspiring. For most people, I imagine it would be a terrifying experience to be amongst falling rocks up here, as Elijah was!

Our Saudi guides gave us time. They understood that this was a deeply spiritual experience for us, as this place was significant to them too. Jewish and Bedouin traditions have continued for centuries to call this peak *"The Mountain of Moses."*

Roland, Lilian and I ate and drank in Elijah's cave. We were quiet, prayerful, reflective, blessed.

Before we started our decent, I remembered my two dozen handkerchiefs that I had brought up to the cave with me. I gathered some sand and stones and placed them in the plastic bag with the anointed handkerchiefs. These I would later give to my friends and family. Then I stood at the opening of the cave and loudly declared the name of Yehovah. Following this, I felt moved to declare the Aaronic blessing in Hebrew out over the vista before me.

***Y'varekhekha Yehovah v'yishmerekha***
Yehovah bless you, and keep you;

***Ya'er Yehovah panav eleikha vichunekha***
Yehovah make His face shine upon you,
and be gracious unto you;

***Yissa Yehovah panav eleikha v'yasem l'kha shalom***
Yehovah lift up His countenance upon you,
and give you shalom.

What an honor, privilege and joy to declare this from His holy mountain in His language after long weeks of learning what had been so difficult for me. And I said every word without faltering! It had not at all been in my mind to do this. Later I would find out that one of our team members who ascended the mountain to the Plateau was also moved to declare the name of Yehovah and to speak out the Aaronic blessing in Hebrew, as well as one of our team members who ascended the mountain to the Peak. All three of us had been moved to do exactly the same!

So in all three places on the mountain that day the Aaronic blessing was declared in the very same language Yehovah first gave this prayer of blessing to Aaron for His people. And in all three places—the Plateau, at the Peak and in Elijah's cave — His name was proclaimed! Just to think.... this was probably the first time Yehovah's name has been shouted from Mount Sinai in 3500 years!

What's more, there was a ram's horn (shofar) safely hidden in the rocks on Mount Sinai's peak. It had been left there by Keith Johnson who had shared with us the co-ordinates. Once found, the blast of the ram's horn was heard all the way to Moses' Altar! This sound was thrilling, but this sound was also significant! Why?

Blowing of the ram's horn, or the *Yovel* as it is called in Hebrew, is an ancient ordinance given to Israel at Mount Sinai, and the law that ordained the observance of what would be called the Yovel. An instrument to mark the changing of times, the beginning and ends of eras.

In Exodus 19 there are two different horns blown at Mount Sinai.

In verse 13, there is the description of the *Yovel* (לבֵי). Yet, in verse 19, there is the shofar (רפָוֹשׁ) blast. Why the *Yovel*? Because the Y*ovel* is the ram's horn blast of the *Jubilee*, and the word *Jubilee* comes from the Hebrew word *Yovel! Jubilee* is the symbol of Yehovah's people being set free from slavery! The sounding of the shofar or ram's horn would signify the end of the forty-ninth year and bring in the fiftieth i.e. the *Yovel* or *Jubilee*. The sound of the ram's horn evokes Yehovah's power and blessings.

"*Not a hand is to touch it, but he shall certainly be stoned or shot with an arrow, whether man or beast, he shall not live. When the Yovel sounds long, let them come near the mountain.*" (Exodus 19:13)

When the women who had remained at Moses' Altar heard the sound of the shofar, they experienced the joy of The Spirit bubbled up within them and they also boldly shouted the name of Yehovah three times!

All those within range, heard the shofar sound out over Har Yehovah for perhaps the first time in thousands of years. What an extraordinary experience this was for us all, no matter where on the mountain we were! And I believe in the spiritual realm too. I imagine that all the heavenly beings and the great cloud of witnesses were tuned in to the sound! And who was it from our team that found and blew the Yovel? Who would be the one divinely designated to do this? It was Danielle, who's husband was at home looking after their son.

And so it was time for Lilian, Roland and I to descend the mountain from Elijah's cave with our Saudi guides. It was hard to leave. I was reluctant and sad to descend. I didn't want to. I felt safe on the mountain in the embrace of my Creator. To be in Elijah's cave was a sanctuary for my soul. I tried to hold on to that feeling of wonder, strength and security.

About a third of the way down, I slipped on loose stones and **fell head over heels down the mountain. I sprained my wrist.** I broke my pole. I gashed my shin. I could only use one arm to finish the difficult descent. I clung to my Saudi guide Def. My other arm was in a makeshift sling. Def was a god-send in getting me down safely.

Our epic journey continued, bringing us to the Yom Suph, actually the Gulf of Aqaba - an arm of the Red Sea - in Haql.

The view of the Red Sea from my third-floor hotel room was as beautiful as it was glorious! This was the water that had been supernaturally split apart to make a path for the Israelites to escape from Pharaoh and his army. Yehovah had done this, *with a strong hand and an outstretched arm, for His mercy endures forever* (Psalm 136:12). The mighty and monumental walls of water, held back for the sake of Yehovah's beloved, were now a visual testimony to me of Yah's faithfulness to deliver. Not just having delivered the Israelites from Pharaoh, but over and over again in my life He had been my great Deliverer too. Now these waters were still, restful and shimmering in the sunshine.

On the morning of March 12, 2023 in that hotel in Haql by the Red

Sea, I had a lucid dream where my spirit had an encounter with Yehovah.

I was asked, ***"Will you marry Miles?"***

I answered, ***"I can't. Things aren't ready."***

Again I was asked, ***"Will you marry Miles?"***

Again I responded, ***"I can't. Things aren't ready."***

A third time I was spoken to. This time I was asked, ***"Will you trust Me?"***

I answered, "***Yes***".

It was 5 am in the morning when Miles knocked on my hotel room door. I was staying on the third floor and Miles was in a room on the first floor. He had heard me call his name and this awakened him from a sound sleep. Miles knew it was something supernatural, as there was no way he could have heard that coming from the third floor. He knew he should come and see if I was alright. Miles got up and got dressed, then came up to my door and knocked. It seemed to me that his knock came just after I had answered "*Yes*" to Yehovah. I quickly got up, wondering who could possibly be knocking at my door so early. It was just after 5 am. It was Miles!

"***I heard you calling my name***", he said to me while standing in the hallway.

I tried to make sense of what he was saying, while aware that our conversation might wake up others in neighboring rooms.

***"No. That's ridiculous! But even if I was… how could you hear me?"***, I loudly whispered.

***"I am on the third floor and you are on the first!"***

"***I did!***", declared Miles.

***"You'd better come in!"*** I said.

It was all so strange, and before I knew it,
I found myself saying to Miles:
***"I have just told Yehovah that I would marry you!"***

# Unexpected

Miles had been shocked but very, very happy when I told him of my agreement with Yehovah to marry him. He had often thought that if I was ever going to marry him, it would take divine intervention, and that is exactly what happened!

Of course, no one expected to have a wedding take place at Mount Sinai, least of all us! When Miles broke the news at the breakfast table that morning, everyone was delighted! I think I was in a daze that day as I do not remember much of it.

We had one evening and one day to prepare. We met that evening with a small group - Pastor Art, who would officiate at the wedding, Janice and David, Jenny Webster, Roland and Lilian, and Patrick. Miles and I shared the supernatural events of the night hours.

We remembered the Ketubah ceremony Yehovah ordained at Mount Sinai (Exodus 19) was a wedding covenant. In fact, the Jewish wedding ceremony is based on those events that took place at Mount Sinai around 3500 years ago. It included a chuppah (covering) to represent the cloud covering and presence of the God of Abraham, Isaac and Jacob, and a Ketubah - the written agreement setting out the terms of the covenant. Janice suggested that this occasion would be the perfect time for the team to stand together before Yehovah and make our own oath of obedience to Him. Our gathered friends asked if Miles and I would be willing to share our Ketubah ceremony with the whole team, representing all those who would be the bride of the Messiah. Without hesitation and in unison Miles and I said, ***"Yes!"***

*"Torah is often referred to as a marriage agreement, and with their acceptance of the commandments they became the 'bride'. This sets in motion the rest of the biblical narrative of how a loving God seeks to restore His wandering bride. And it is through the renewing of this covenant in the death, burial, and resurrection of Yeshua that we have finally found peace and restoration with Yehovah. And I thank Yah that this covenant is no longer written on tables of stone, but now it is written on our hearts."*

And so, the wedding planning began, starting with Jenny, Janice and I earnestly praying that Yehovah would provide a cloud as a chuppah (cover) – just as He had so many centuries before for the children of Israel in the wilderness, when He spoke to Israel from the mountain. [The Saudi Arabian government severely frowns on bringing anything Jewish / Hebraic into the country or to the mountain, so no one had brought with them a prayer shawl, that could otherwise have been a bridal cover for me.]

The day before the wedding, would be the team's second day at Mount Sinai. With this being the only one day to prepare for all matters of marriage, I decided not to go but stay at the hotel. This was the decision of Pastor Art, Janice and Jenny too. So while the other members of the expedition were climbing rocks, finding inscriptions, taking photos and filming, Janice and Jenny, busied themselves writing out individual copies of the Ketubah – a written agreement - for us each of our team. It was tremendous labor of love and commitment, as each handwritten copy took a good length of time! In the Ketubah, they carefully outlined the information found in Exodus and Deuteronomy about the covenant and Yehovah's promises to those who obeyed from the heart.

Pastor Art and I met for some pre-wedding counsel, then while he prepared his thoughts and notes for the ceremony, I was faced with the big question every girl has for her wedding day… What was I going to wear? My options were limited. In fact, did I even have any options other than my climbing clothes? Ah… no. Yet even if I had, it would not have been permitted to wear any bridal attire at Mount Sinai in Saudi Arabia.

For the remainder of my day, I prayed and tried to rest. The antidote to the nervousness I felt, was to revise my conversation with Yehovah - "***Will you trust Me?***", and my response, "***Yes***".

The evening meal for our team was at the best local restaurant in town, where we had eaten a number of meals together. Our last meal here was the evening after climbing the mountain. Our conversations that night were energized and animated as we shared our varying and incredible experiences of that day. And now tonight, at our pre wedding evening meal, we were once again in high spirits as we anticipated tomorrow's ceremony on Mount Sinai - the Mountain of God. All would attend. Tomorrow, I would become Mrs. Catherine Jones!

It so happened, that in the same restaurant that evening was Pastor Steve Cioccolanti from Melbourne. He and his wife had just arrived to visit Mount Sinai. Having heard that Dr. Jones and the team were in the restaurant, Pastor Steve came to our table during dinner to chat and exchange contact details. Tomorrow, he would be on the other side of the mountain at the Split Rock while we would be at Moses' Altar.

# The Wedding

***"...be ready on the third day"*** (Exodus 19:11)

On March 14th 2023, I woke in my Haql hotel room realizing the magnitude of the day that was before me. I was getting married, when just two days prior it wasn't even a thought in my mind. And, we were about to be married on Mount Sinai at Moses' Altar, beneath the cleft in the rock and Elijah's Cave. How does one process that?

With a badly sprained wrist, I showered and washed my hair as best as I could. This was not easy, or fast! In hindsight, I should have asked one of my sisters to at least help me get my clothes on, dry my hair, and assist me with packing. We would make our final departure from the hotel after breakfast.

I dressed in my climbing khaki pants, white linen shirt and climbing boots. On my head I would wear a wine colored scarf that Lilian would help me with before we got to Moses' Altar. I would also -ear the Esther belt. It had dawned on me yesterday why Renae had been led by the Ruach Hakodesh (Holy Spirit) to have me bring this Esther belt to Saudi Arabia. What had she said to me?

***"There is a special purpose for you with this. I don't know what it is, but you will find out."***

**Surely**, it was *for such a time as this, I thought!* I was getting married to Miles and **we were all going to renew our covenant vows of marriage to Yehovah, The King, at His Holy Mountain!** My heart was filling with quiet wonder at what was about to take place.

Breakfast was once again served to the team in the lobby of hotel. I had no appetite but I forced myself to eat something small. As we left in our convoy of Land Cruisers, the sun shone bright in a clear blue sky, and just like every other day, it was hot and dry. Once more, we were making the three-and-a-half-hour journey to the mountain. This time, I traveled separately from Miles. I was in a Land Cruiser with Janice, who would be my matron of honor, and Lilian, who would help with my headscarf. Miles traveled with Roland, who would be his best man, and Patrick his video producer. This was the last day for us to be at Mount Sinai. It was Tuesday.

Miles was also wearing his pants, shirt and boots which were his daily expedition wear throughout our days in Saudi Arabia. Today though, for the wedding, he had decided not to wear his Indiana Jones hat. Later he regretted this as his face was badly sunburned.

Our convoy of Land Cruisers arrived at the mountain. Since this was the last day we would be on this research trip, we had asked if our Saudi drivers would give us privacy for our final gathering together at the Mountain. We had all been careful to not inform them that we were performing a wedding ceremony, as we did not want to have this stopped or interrupted. Our drivers therefore stayed with the Land Cruisers.

Lilian helped me to fix my head scarf and place the Esther belt on me. The sash's soft wine color matched my headscarf. Lilian had wrapped the fabric twice around my waist, and the bell ends hung low on the side of my body. The delicate sound of the small gold bells made me feel feminine and beautiful for my wedding day! Could there be any better reason for wearing the Esther belt than this day, I wondered. But of course. This is what it was for - a wedding at Sinai!

Our expedition team made our way to Moses' Altar. As we did, some of the team foraged for wild grasses that would be my wedding bouquet. It was to be a perfect wedding bouquet!

At Moses' Altar with the twelve pillars that had represented the twelve tribes of Israel, we set up cameras and phones, and found the best places to stand and film. For me, and I'm sure for others too, I felt a holy reverence and awe in my heart for what we were about to be a part of. It was a quiet expectation. We had even decided beforehand, it was only right that we would stand before Yehovah at His Holy Mountain with our shoes off. Without a doubt, we were on holy ground! There was great joy in our hearts and on our faces.

For some reason, not all of our team had made it to where we were at Moses' Altar. What had happened to them, we wondered. Instead of our full head count, there was only Miles and I, Pastor Art, Janice, Roland

and twelve and others. Really? We had exactly 12 witnesses, correlating to the remnants of the twelve pillars of the twelve tribes of Israel where we now stood. This of course had not been planned. We were to later reflect, that it is no light thing to serve the Creator of the Universe, who is the Elohim of Abraham, Isaac, Jacob, Moses and Yeshua. Of the thirty-plus original tour members only fifteen took part. Half went home a week earlier and others missed this wedding ceremony for mysterious or even inexplicable reasons.

Wide-eyed at the thought of 12 witnesses, someone then noticed a cloud hovering near the top of Mount Sinai. We were even more excited! Yehovah had heard our request for a chuppah, a bridal cover.

Miles and I, along with Janice and Roland either side of us, positioned ourselves. Pastor Art began the wedding ceremony. I was somewhat nervous, but the more overwhelming emotion was definitely wonder and awe. Was this really happening? I was just a woman from New Zealand, who just happened to flee domestic violence and find a safe house in Kerrville, Texas, who just happened to meet Dr. Miles Jones at a restaurant one night, who just happened to be a biblical archeologist and researcher, who just happened to be leading an expedition to Mount Sinai, who just happened to be standing before me now on Mount Sinai before twelve witnesses, who just happened to be speaking his wedding covenant vows to me, while there just happened to be one solitary cloud over our heads!

I remembered my dream two months prior where Yehovah spoke to my heart saying, "***There is more ahead of you than time would suggest Fiona.***" Wasn't that the truth! Since that time, I had been saying, "*Hineini. Here I am. I am ready Yehovah.*" And now, *here I was*, before The King! There was no 'pause' button I could press to let this all sink in.

We had prepared our hearts and our minds to receive what Yehovah

had in store for us. Though none of us had anticipated a wedding and a covenant ceremony with Yehovah, we had begun our preparations long before we arrived, in fasting, and prayer and diligently seeking to better know Yehovah and Yeshua in His Word and through prayer. Some of us had fasted and prayed for three days in preparation, as Esther did before appearing before the King. After all, our first visit to Yehovah's Holy Mountain had coincided with Purim!

And now, as Miles lovingly nudged me, I once again heard the delicate sound of the bells on the Esther belt. What was it Renae had said about the belt?

*"It's worn to bring the essence of Yah's presence where He desires dominion.*
*It's a tangible visual, that changes the atmosphere through His Presence and His Directives.*
*It is only to be worn when He directs, and for His purposes."*

I believe the Spirit of Esther ***- to rise up and do what you are made to do in the name of Yehovah -*** was with me there at Mount Sinai. Yehovah had given Esther such a heart, to rise up and do what He made her to do in the name of Yehovah! Now, the Spirit of Yehovah was rising up within me to accomplish and fulfill His plan and purpose. I too was born for such a time as this.

Miles and I recited our solemn vows to each other in the presence of Yehovah, Yeshua and Israel i.e. our brothers and sisters present, who like us were grateful ingrafts to Yehovah's covenant people through faith in Yeshua. We too, are now called the children of Israel. And so we were pronounced husband and wife. I was now Mrs. Jones. In such circumstances, shouldn't my name have been something more grand than Mrs. Jones?

Pastor Art turned to address all of us who were humbly gathered, to begin the second part of the covenant ceremony. He led us all in confession of our sins, repentance, and our asking Yehovah's forgiveness, prior to partaking of the communion of crackers and juice. This we did in remembrance of Yeshua giving His life on our behalf, redeeming us by His own blood (Luke 22:19-20).

[We were later to wonder if this covenant meal, in remembrance of Yeshua, broke the curse of the golden calf worship. Exodus 32]

*"See, the days are coming," declares Yehovah, "when I shall make a renewed covenant with the house of Israel and with the house of Judah, not like the covenant I made with their fathers in the day when I strengthened their hand to bring them out of the land of Mitsrayim, My covenant which they broke, though I was a husband to them," declares Yehovah.*
*"For this is the covenant I shall make with the house of Israel after those days, declares Yehovah: I shall put My Torah in their inward parts, and write it on their hearts. And I shall be their Elohim, and they shall be My people.* (Jeremiah 31:31-34)

For many centuries, Jewish weddings took place on the third day of the week, to remember that they prepared for three days to meet Yehovah at the Mountain. They marry beneath a canopy – called a chuppah – in remembrance of the covering of clouds at Mount Sinai. On the third day, as part of the covenant ceremony, Yehovah spoke to the people from the mountain, beginning with the ten commandments. (Exodus 20:1-17)

## The Written Ketubah

Pastor Art turned to the people gathered at the altar and read from Exodus, and as part of that ceremony, we committed to do all that Yehovah told us to do. Three times we proclaimed, "***All that Yehovah has said, we shall do***." This echoed the momentous occasion of the covenant first presented over 3500 years ago to the children of Israel in Exodus 19:5-6.

*So Moses came and called for the elders of the people, and laid before them all these words which Yehovah commanded him.*

**Then all the people gathered that day answered and said, "All that Yehovah has spoken we will do."** *So Moses came and told the people all the words of Yehovah and all the judgments.* (Exodus 19:7)

***Once more, those who gathered there said with one voice, "All the words which Yehovah has said we will do."*** *Then he (Moses) took the Book of the Covenant and read in the hearing of the people.* (Exodus 24:3)

***A third time they said, "All that Yehovah has said we will do, and be obedient."*** (Exodus 24:7)

This was the first covenant to be made at The Mountain of Elohim since the time Moses and the children of Israel came together. We, now also children of Israel, once again said ***three "I do's"***, and so renewing our wedding vows with Yehovah. Even more significant, was that we were the first New Covenant believers who had made a Ketubah with Yehovah at Mt. Sinai!

At the conclusion of the ceremony, we signed our Ketubahs. Each person requested signatures from the others. It was wonderful to see the zeal to complete the Ketubah witness! As we were all about to leave this incredible place and experience, I asked if I could speak the Aaronic Blessing over everyone present - and in Hebrew too, of course. As I did, I spoke with a boldness I had not possessed before.

With our hearts full to overflowing with praise and thanksgiving, we gathered up everything, and now for the last time we made our way back to our Land Cruisers where our Saudi drivers were taking a siesta. We had no idea that we were about to literally see how Yehovah *changes the atmosphere through His Presence and His Directives*!

*In the third month after the children of Israel had gone out of the land of Egypt, on the same day, they came ... and camped there before the mountain. And Moses went up to Elohim, and Yehovah called to him from the mountain, saying, "Thus you shall say to the house of Jacob, and tell the children of Israel: 'You have seen what I did to the Egyptians, and how I bore you on eagles' wings and brought you to Myself.*

*'Now therefore, if you will indeed obey My voice and keep My covenant, then you shall be a special treasure to Me above all people; for all the earth is Mine. And you shall be to Me a kingdom of priests and a holy nation... And Yehovah said to Moses, "Behold, I come to you in the thick cloud, that the people may hear when I speak with you, and believe you forever."*

*Then it came to pass on the third day, in the morning, that there were thunderings and lightnings, and a thick cloud on the mountain;... Now Mount Sinai was completely in smoke, because Yehovah descended upon it in fire. Its smoke ascended like the smoke of a furnace...*
(Exodus 19:1-9,16,18)

Once back to the Land Cruisers, our Saudi guides brought out food. We sat on the surrounding scattered rocks to eat. Miles and I were eating our first meal together as husband and wife. Even this was to prove memorable beyond belief.

Heavy thunderclouds were starting to replace the small puffy white cloud, as we walked back from Moses' Altar. But then we had barely been able to finish eating our meal when it began to sprinkle and a powerful wind arose, blowing dark clouds into the area. We felt a sprinkle of rain, as we began to see ever darkening thunder clouds moving in toward the blackened peak. The wind took Miles' hat off and threatened to do the same with my plate, but I clutched it tighter in my hands trying to eat what I could of my meal. But that was not to happen either - the great wind blew ever harder!

I was so enthralled with what I was seeing, hearing and feeling, that I ran to Janice and began to sing, "*How great is Yehovah (our God), sing with me how great is Yehovah (our God), and all will see how great, how great is Yehovah (our God).*" Janice was quick to join me in praise. We were truly amazed, and as we sang our hearts out together to Yehovah, while at the same time we each were videoing all that was before us.

The dark thunder clouds were now completely covering the top

of Mount Sinai – just as the clouds had covered it when Yehovah spoke from the cloud to the people below, proclaiming the covenant to them. Yehovah was giving a clear sign that He heard our voices! But He wanted to make sure there was no doubt. His celestial miracles were only beginning.

Lightning and thunder came closer and closer. The clouds grew darker and covered the blackened top of Mount Sinai! We KNEW this could not be a coincidence. Then came the rain! It was then we started scrambling to get into the Land Cruisers. The rain became a deluge! It was the kind of torrent that the windshield wipers can not keep up with and it did not relent.

We were rejoicing at this miracle! We were in a desert that receives less than an inch of rain each year. The water fell so hard and so fast that even after getting into our Land Cruisers, the drivers could barely see! When we thought it could get no worse, it began to hail and the ground turned white with the marble-sized hail!

Miles and I were in our own Land Cruiser with a driver. We were, after all, the newlyweds! I was overwhelmed not only by these celestial weather events that we were witnessing, but that I was now Mrs. Jones living a grand adventure in the wilderness of Sinai!

Our drivers had difficulty dealing with the conditions. They desperately wanted to get out of there before getting hopelessly mired in the flooded roadway which had been graded but not yet paved at this stage. The day had blackened like nightfall as the storm clouds now covered everything to the horizon. This weather was something extreme and unexpected for them to be dealing with.

We were all exuberant as we drove to *the camp*. Still, Yehovah was not finished yet. It began to hail. For as far as the eye could see, small, white balls of hail covered the earth – looking as if the ground was covered in manna!

Yehovah had already prepared the footsteps of each of our lives to bring us to this moment in time. He knew every single one of them and His timing was precise and calculated. There was yet another example of just how precise His timing was. Janice's husband David asked her, *"What day is this?"* It was Tuesday. The third day of the week. Then he asked, *"How many times have we been here?"* Janice said, *"This is the third day!"*

We came to realize another wonder-filled thing, that not only was it the third occasion our team had been at Mount Sinai, but we were at Mount Sinai renewing our vows on the **third day** of the week - as per Exodus 19:11! As a matter of fact, we were in the camp of the children of Israel! The only thing that would have made this event more powerful was to hear the voice of Yehovah! But the ever louder booming of thunder and the bright crackling of lightning all around us, was this not the voice of Yehovah? Had we not heard the voice of the shofar on the mountain as well?

***And it came to be, on the third day in the morning, that there were thunders and lightnings, and a thick cloud on the mountain. And a voice of a shofar was very strong, and all the people who were in the camp trembled.***
(Numbers 19:16)

At this point, Janice and I were frantically trying to call each other from our respective cars as we knew that Yehovah was performing a MASSIVE MIRACLE for us. But our cell phone service had failed! So when our cars aligned, we put our windows down and shouted to each other above the noise of the hail, "***HE heard us! HE heard us!***"

We were all laughing and marveling at what we were experiencing and seeing. A great testimony of Yehovah moving in his people. Was it possible that ancient portal of Psalm 24 could manifest in modern times? Psalm 24 did call it an *eternal portal* opened upon Har Yehovah through which the Almighty does come through as He chooses. As if our minds weren't blown enough! We came down from The Mountain and danced with the Arabs in the rest area beside the road. Everyone was astounded. It almost never rains like this in Arabia.

Later I was to reflect on the dream I had February 14, 2023 where a selection of different shades of blue linens were put before me and I was to choose one, and then seeing the word ISAIAH. I wondered if the deluge of rain was to be symbolic of a Mikva, or baptism. That we, His bride, were being purified.

***A highway shall be there and a road, and it shall be called the Highway of Holiness. The unclean shall not pass over it, but it shall be for others.*** (Isaiah 35:8)

To go home and tell the story of what happened today may not be believed, except that that we had filmed and photographed everything of the ceremony and of these extreme celestial weather events. And now, as our time in Saudi Arabia and at Har Yehovah was drawing to a close, we headed back to Tabuk for our final night, and this brought about a change in plan.

The Millennium Hotel, the hotel we had stayed in on our arrival to Saudi Arabia and were to return to now for our final night, was no longer available to us due to the weather events. This was because the government's thousands of Neom workers had to abandon their work and seek shelter. All of the shifts of workers required housing till there could be a return to work. The government seconded the Millennium Hotel, which meant we had to find an alternative hotel. This was to be a blessing in disguise too, as another hotel was found that was even nicer. Well at least for Miles and I it was, as we were given the honeymoon suite, complete with a Jacuzzi! Had someone other than Yehovah organized this for us? It didn't appear so. No one seemed to have any idea how it happened! At least no one confessed.

The great storm from Mount Sinai followed us to Tabuk, where we drove through flooded streets. Yet as the storm raged outside our hotel, Miles and I were unaware. I am thinking we were excused for having other things on our mind. And it had been a momentous day!

At 3am the next morning we were loading our luggage into the Land Cruisers for the final time. From Tabuk we made the hour and a half flight to the airport in Jeddah, from where we caught another flight back to the States. Having boarded the plane in Tabuk to fly to Jeddah, a Saudi Arabian man shared photos and video of the snow that fell at Mount Sinai, Jabal al-Lawz only hours before. That had started soon after we left. Yehovah continued with great wonders! A deluge of rain, hail and now snow! Yehovah continued to proclaim through the storm that he heard his children's vows.

## We were soon to discover other things once back home:

On the day of our marriage at Sinai, March 14, 2023, in the south of Arabia, there was a monumental flood. That weather started in the north of Arabia with the Ketubah ceremony at Sinai. Remember that

Saudi Arabia was entering into summer at that time, temperatures above 100 degrees Fahrenheit. That year, during the first weeks of the summer season there was flooding, hail and heavy snow across the region. In Tabuk, Riyadh and Jeddah they were digging out of the snow. There was incredible flooding over much, if not most, of Arabia and Israel. There was flooding in Oman. There was flooding in Yemen. There was flooding in the Negev Desert of Israel. Rivers of snow and ice water began flooding through the cities of Arabia!

Three weeks before our arrival, a geyser had erupted in the Rub al Khali in the south. It created an incredible flowing river that gathered into a large lake in Wadi Rum. There were people riding across this huge lake on jet skis. In Isaiah 43:19 it says, "*I will even build a road through the wilderness, and rivers in the desert.*

It was intriguing to learn that at the very same time we were at Mount Sinai when the rains, floods, and hail began, Pastor Steve Cioccolanti and his group were on the other side of the mountain, at Split Rock! He recorded video of what they saw and experienced of the extreme weather events.

And as you can imagine, these extreme weather events captured a great deal of attention around the world. In fact, members of our team discovered that the typical rainfall in the Tabuk Region of Saudi Arabia is approximately 33.5mm (1.32″) each year. It appears the average yearly number was exceeded in one single hour on March 14, 2023, the day of the covenant ceremony! In the south of Arabia, massive amounts of rain fell within a few hours. Roads and bridges were washed away. The streets became rivers and washed away even the large trucks that tried to cross them.

Little did we know that these extreme weather events would

continue for months and spread to all parts of Arabia and the entire region. Yehovah had got the attention of every man, woman, and child, in Saudi Arabia. We started to wonder if the storm could be symbolic of Yehovah cleansing us? Would it be a biblical symbol of the outpouring of the Ruach as described in the book of Joel?

The outpouring of rain as given in Scripture is given as a symbol of the outpouring of Yehovah's Spirit. Even a year later it is reported that the sands of Arabia where we were, are covered with green grass.

Isaiah 35:7,8 and Isaiah 41:18,20 are fulfilled. Isaiah 43:19-21 and 44:1-4 are both in the midst of happening. In the past months fountains are springing forth in the Desert of Saudi Arabia. Water flows once more at Mount Sinai. The almond trees on Mount Sinai have blossomed. The Euphrates is nearly dry. These are incredible and awe inspiring times we are living in! The call is to repent and return. To Teshuvah. As Yehovah called Israel to return in its last days before judgement, He is now calling all people. All nations.

# Epilogue

*"It is hard for our modern minds to comprehend the possibility that behind remarkable events in the world, could lie a deeper story, of an unseen Hand.*
*Remove any of the events leading up to them, the twists and the quirks, or move any of them out of place, and none of it would have come out exactly as it did.*
*Had we not... had I not... had they not... had this not...*
*And yet it is not so much that the mystery works inside the parameters of human history: it is that human history works inside the parameters of the mystery."*

**- Jonathan Cahn - *Josiah Manifesto***

This has been an incredible story to write. A mystery to unveil. A putting together of the pieces. It could be considered by some to just be a random lot of events and happenings that all strangely lined up. However, I believe nothing was of happenstance or accidental, and that Yehovah's plans and purposes were brought together in a spectacular way. He lined all things up and provided numerous '*signature cards*'. What Yehovah did at Mount Sinai was with phenomenal prowess!

***A man's heart plans his way, but Yehovah directs his steps.***
(Proverbs 16:9)

Of all the people in the world Yehovah could have chosen to be married at His mountain, He chose Miles and me. Why? We know that Yehovah spoke to ancient Israel through prophetic acts. Not all the prophetic acts were understood at the time. Perhaps all that happened to us was, or is, to happen without any knowledge of the prophetic significance on our part. It is simply enough that it was Yehovah's purpose.

Yet I am wondering if the experience I had at Har Yehovah might be to unravel a mystery that would bring not only understanding and healing to my own heart, but understanding and healing to the hearts of many.

On the morning of February 29, 2024, Renae called me with great excitement in her voice. A revelation had risen up in her that morning as she had come out of her prayer closet and was walking across her kitchen floor. The revelation was this, that in all the world, Yehovah chose me, He ordained me to be there at His holy Mountain. That He wanted to marry me there. Yes, I had married Miles, and yes those of our group also pledged and signed the Ketubah, but He had chosen me to be a bride.

After all the devastation in my life, Yehovah showed me that my life counts. It was as if Yehovah said to me, "*I'll tell you who you are — I'll put you on My holy mountain as a bride! - no longer wounded, abandoned, rejected. I am your husband!*" Mount Sinai was to heal my heart. It was to tell me - let go of the old me. Don't be bound by your past. Let it go. Will you be Mine? Free of everything that has gone before.

This is the biblical mystery of the bride and groom! In the beginning was Elohim, and Elohim is love. And so before there was anything else, there was love. We were *made* for love! The events that played out was a love story between Yehovah and me. A phenomenal love story! What a romance this was, on March 14, 2023 for the Kingdom of Elohim. I should never doubt it. I am of the bridegroom's house and I am in covenant! He is my story from beginning to end. He is the author of my faith and the finisher of my story. All my life experiences are in Yehovah's hands. As Renae said to me, "*this is the time for you to get on the horse and ride with Him!*"

Our Ketubah ceremony at Mount Sinai was the first, since Moses. Perhaps the last! The Saudis will now zealously guard the groups at the mountain to make sure such religious ceremonies do not happen. Our Sovereign Elohim brought us to His Mountain as the last expedition to cross the desert in our Land Cruisers. In the 10 days we were there, they were building the nearby city of Neom so fast that we had gone to The Mountain across the desert, and were leaving The Mountain that day on paved roads. Everything from that point forward would be tightly regulated, even though many more people would be allowed to visit The Mountain.

It turned out that our agenda for the trip was one thing, Yehovah's purpose was another. We were in the right place at the right time doing what He had called us to do… "*for such a time as this" (Esther 4:14)*. We were on Yehovah's program. Truly amazing. Such a coup was Yehovah's planning.

All that happened on our journey to Mount Sinai was above and beyond all I could have imagined and hoped for, and the telling of this story has been for me a mystery, a joy, a challenge, and it is my testimony. Yehovah is doing a new thing and miracles continue at Mount Sinai. I believe Yehovah is calling His people to Teshuvah - to repent and turn back to Him - and determine to do what He asks us to do, because of His very great love for us.

# Some final words from my beloved, Dr. Miles Jones:

*"Now it shall come to pass in the latter days*
*That Har Yehovah's (the mountain of the LORD's) house (of prayer)*
*Shall be established on the top of the mountains,*
*And shall be exalted above the hills;*
*And all nations shall flow to it.*

*Many people shall come and say,*
*"Come, and let us go up to Har Yehovah (the mountain of the LORD)*
*To the house (of prayer) of the God of Jacob;*
*He will teach us His ways,*
*And we shall walk in His paths."*
*For out of Zion shall go forth the Law,*
*And the Word of the Lord from Jerusalem.*
(Isaiah 2: 2-3)

**We are seeing prophecy being fulfilled!** "*Behold! I will do a new thing. Now it shall spring forth. Shall you not know it? I will even make a road in the wilderness, and rivers in the desert!*" (Isaiah 43:19) We drove out of Mount Sinai on a newly paved road. We were perhaps the first to drive it except for the workers who made it. The river that erupted in the Rub al-Khali is not the last. The massive flooding has turned dry wadis (stream beds) into rivers all over the Saudi Peninsula.

"*Har Yehovah's house of prayer shall be established on the top of the mountains... and all nations shall flow to it!*" (Isaiah 2:2).

The Mountain of God is a Temple, a house of prayer. It has been hidden for 3500 years. Now it is open again. We were not the only group on the mountain. People of all nations have already started to come. Soon there will be thousands of them coming in on tourist buses daily from the hotels built nearby on the main highway going to the new city of Neom. Millions will come every year from all the nations of the world. Israelis have been to the mountain already. They will come from everywhere. Now that we know where the hidden mountain is located!

"*You are those who forsake Yehovah, who forget my holy mountain!*" (Isaiah 65:11).

Those days are now over, the covenant on the mountain has been renewed, not only by our small group - but soon by millions of believers. The Mountain of God has even been hidden in Scripture. There are various names for the mountain; Har Horeb, Har Sinai, Har Elohim but Har Yehovah - the true name of God - has been hidden under translators' conventions. It is variously referred to as the Mount of the LORD, or the mountain of the LORD, or the hill of the LORD - all are written as *Har Yehovah* in the Hebrew Bible. Therefore, verses like Psalms 24 and Isaiah 2:3 and various others are obscure to the reader who knows not this is the Mountain of God - Har Yehovah.

"*Come let us go up to Har Yehovah... and we shall walk in His paths. And out of Zion shall go forth the Law, and the Word of the Lord from Jerusalem!*" (Isaiah 2:3).

Haven't those things already happened? Remember these are end times prophecies. The hidden is coming to light. Other hidden things in the Law of Torah will be revealed. Now that I have recovered, authenticated and begun the translation of *The Hebrew Gospels* the *"Word of the Lord"* (Yeshua) will be available from the original Hebrew manuscript tradition. There is a massive movement of the Spirit coming. HalleluYAH !

# Appendix A

## In Memoriam:
## Elizabeth Chesley Baity

**Elizabeth Chesley Baity 1907-1989**

**We know Elizabeth Chesley Baity as a pioneer in the field of Archaeoastronomy.** She gave it its name. Her 1973 article, "*Archaeoastronomy... So Far*" in Current Anthropology magazine set the cornerstone of the field. She wrote eight articles on Archaeoastronomy in the early seventies - turning the new science into a going concern. By the early eighties, her vision had become a reality. Celestial events sufficiently recorded in ancient records could now be retro-calculated by astronomical computer software. This has revolutionized the study of chronology giving us anchor events in ancient timelines that we know to be certain. It would be hard to overestimate the import of this new science in the study of ancient history!

We are still on the precipice of that revolution in historical revelation. Slowly, but surely, the new science of Archaeoastronomy has proven itself - but historians have yet to give up their treasured orthodox timelines based on historical guesstimation. The New Chronology will eventually change that. "*A new broom sweeps clean*" - it is said - but it usually takes a generational change. That change will raise Elizabeth Chesley Baity to the prominent historical place which she certainly deserves. In the meantime, I would like to share some of my knowledge about **this remarkable scientist, anthropologist, writer, world traveler, painter, poet, woman, wife and mother.**

**I met Elizabeth Chesley Baity when I was far too young to care.**

I was about five years old and totally unimpressed with the rows of books on her many bookshelves. I was far more interested in how to get into the atrium in the middle of her home. She and my mother were talking very animatedly. I have often wished to return to that moment and sit silently with them, recording their conversation in memory. Elizabeth had just won the Newbery Prize for her first book, *Americans before Columbus*. My mother and father were also writers, their patient tutelage was the primary impetus in becoming a writer myself.

When I first became interested in ancient inscriptions, my mother gave me Baity's book, *Americans before Columbus*, to aide in my research. I eventually earned three degrees in languages and linguistics as a result of that early interest. Elizabeth Chesley Baity got her doctorate in Anthropology in 1968 when she was 51 years old. She had married Dr. Herman Baity in 1930, who was to become a Director of the World Health Organization. By the time she began her professional career she had already raised two sons and lived throughout the world with her husband. She spoke, and wrote articles in, Spanish. She published this poem in Harper's Magazine, in 1961.

## Merchant of Rio
### by Elizabeth Chesley Baity

We had seen, wandering down the zebra-striped walk
of Our Lady of Copacabana, where lovers stroll
through the warm endless afternoon, in idle talk,
his great house sprawled behind its guardian wall;
he had been told, you said, by some sharp-witted seer,
doubtless with a lover in the building trade,
that when his house was finished, he would die.

So, for two decades he had made
gallery on gallery, useless tier on tier,
endlessly building, destroying, building again,
yet in the end although the hammers rang
like drum-beats to the saws' whining song,
Death found him out, in that unfinished place.

We smiled and passed, scorning such foolish ways,
and you moved on from girl to girl, and I
never quite finished the poems I had made.

Not that either of us was in the least afraid.

Having studied and written poetry myself, I can see the excellence of craft in her work, "*yet in the end although the hammers rang like drum-beats to the saws' whining song, Death found him out in that unfinished place.*" It was an excellence I suspect she applied to all the endeavors of her life. The poem begins with the word "*we*" and ends with her companion who "*moved on from girl to girl.*" Who was this mystery muse? Was it her husband, a lover, a platonic friend, or perhaps a figment of her imagination? I am not sure which option would be more intriguing - a brief glimpse into the mind of an intriguing woman!

My mother and Elizabeth were a generation apart but knew each other well - as extended family tend to - growing up in the small Texas town of Hamilton. Elizabeth gives credit to her father, Hervey Chesley for much of her motivation and skills in life. I knew Hervey better, since we made family visits to Hamilton from time to time. He had the invaluable ability to make a five-year-old boy, who wandered into his study, feel welcome. He had lots of books as well, somewhat of a savant I understand. I was far more interested in the pipe and tobacco by his chair. He explained it was '*Prince Albert in a Can,*' once Albert married Queen Victoria of England, they began printing "*now king*" on the can, a fact I have never forgotten to this day.

Hervey Chesley's name is on my family tree which hangs in my hallway. I think of Elizabeth Chesley Baity as my great aunt, although I am not sure what the exact familial title should be. One of Elizabeth's paintings hangs on my study wall, part of the flotsam and jetsam of extended family inheritance. All of Aunt Elizabeth's excellent books are on my own extensive bookshelves. For someone I barely knew, she seems to have surrounded me with her painting, poetry, research, and writing - even haunting my childhood memories. I suppose one could say that I have followed in her footsteps.

Of one thing there is no doubt - the work, and the life, of Elizabeth Chesley Baity have shined a great light! Perhaps for each one of us there is no better *memoriam* than this - deeply wish I had had the chance to know her better!

# Appendix B

Review of Books
By Dr. Miles R. Jones

*The Writing of God* presents proof of the seminal event of the Old Testament, the Sinai Covenant, when God handed down to Moses the Word and the "*writing of God*" (Exodus 32:16). The location of the real Mount Sinai has long been debated. Scripture states it to be in Midian in Arabia where researchers have found stunning archeological evidence of the events of the Exodus, including inscriptions from the base of Mount Sinai.

These inscriptions reveal an incredible secret of the Bible. The "*writing of God*" written by "*the finger of God*" (Ex 31:18) is the first alphabetic writing! The Sinai Covenant was an educational covenant calling the Israelites to read, write, and "*diligently*" educate their children.

The evidence taken from the latest linguistic and archaeological science reveals the origin of the alphabet, the context of writing in the ancient world of 2nd millennium BC, and the importance of the Old Testament as an historically accurate source.

*The Writing of God* documents the decline of modern education, especially literacy, as a result of abandoning the moral and the alphabetic principle of God in our schools. The call from Sinai is the spiritual clarion call for today's home-schoolers. To order your copy go to *writingofgod.com*.

***The Writing of God*** is the fascinating culmination of years of research by Dr. Miles Jones. His book on the writing covering the rocks at Mount Sinai in Arabia is nothing short of profound. Backed up by definitive evidence, it is an ancient puzzle coming together in our day. This is a must read!

**Jim & Penny Caldwell**
Split Rock Research Foundation authors *God of the Mountain*

Miles Jones tackles a vital facet of New Chronology research – a revised timeline for the development of the first alphabetic script, from its beginnings in Egypt to the earliest scratchings of Proto-Semitic on desert rocks, through to Proto-Hebrew, Phoenician and Greek. The invention of our ancient alphabet – the alphabet used to write the Ten Commandments and the Book of Genesis – is a fascinating story with major implications for world history!

**David Rohl**, author of *Pharaohs and Kings*,
*From Eden to Exile*, *Lords of Avaris*, and *Exodus*

Dr. Miles Jones is one of the few language experts exploring the link between the origins of the alphabet and the early writing of the Bible. This investigation for new evidence is both controversial and fascinating. The numerous connections brought to light in this new book demand further consideration by all thinkers.

**Timothy Mahoney**, documentary film maker,
*Patterns of Evidence - The Exodus*

*"The stones will cry out!"* Dr. Miles Jones has taken the inscriptions at Mount Sinai in Midian and made the stones speak! ***The Writing of God*** **clarifies** where and when God came to earth and what He said to the world. I cannot wait for your book to be published!

**Dr. Sung Hak Kim**, author of *The Burning Bush*

This is the book that ties it all together! The discovery of the real Mount Sinai in Arabia makes real God's deed - to Abraham's seed - of the Promised Land. Here in Israel, "***The Writing of*** *God*" and the mapping of the footprints of the Israelites are the hope of this generation!

**Michael Rood**, author of *The Chronological Gospels*

# Sons of Zion versus Sons of Greece:

**Miles R. Jones**
Published 2019 order online at
WritingOfGod.com

In the turbulent times of Medieval Spain, a community of converted Jews dared to restore the Messianic Church of the first century - when Yeshua Ha Mashiach and his Apostles still lived. How do we know this? We have their Hebrew Gospels!

The Spanish Inquisition was begun in 1478 specifically to crush the Messianic Church in Spain. In 1492, King Ferdinand & Queen Isabella completed the Reconquista, removing the last stronghold of Islam at Granada. Immediately they turned to the Jews to expel them from Spain and seize their assets. On the deadline of August 31, 1492 - Columbus' three ships left Spain for the New World filled with Jews fleeing the Edict.

Messianics fleeing the Inquisition took the Hebrew Gospels from Catalonia with them. These Gospels ended up in the Vatican Library bound beneath another Hebrew manuscript, undiscovered for 500 years. .Dr. Jones' latest book, Sons of Zion vs Sons of Greece, is not only about that spiritual struggle but also the untold story of the survival of the Hebrew Gospels. The Greek Church divorced itself from the Messianic

**Dr. Jones doing field work in Arabia 1990.**

Church in the fourth century. The GrecoRoman Church then claimed the mantel of the chosen people, a doctrine called Replacement Theology, and proceeded to hunt Messianics and Jewish Christians to extinction. Over the millennia as many as one million believers were martyred during the doctrine wars. It was thought the Messianic Church of Yeshua the Messiah was eliminated along with the Hebrew Gospels. Here is the story of their survival!

## First page of Sons of Zion vs Sons of Greece:

# Chapter One

## The Covenant at Har Yehovah

*And they departed from Har Yehovah*
*[the mount of the Lord] three day's journey:*
*and the ark of the Covenant of Yehovah*
*went before them...* (Numbers 10:33)

**The history of the word of God is, to a great extent, the story of the war of the "*sons of Zion against the sons of Greece*" (Zech 9:13).** The "*sons of Zion*" received a mysterious power as a result of the covenant at Har Yehovah[2], more often referred to as Mount Sinai. That power was passed on, in some degree, to the Greeks. In Genesis, Yehovah is already prophesying to Abraham he will give the Israelites a "*promised land*" now held by, among others, the Kadmonites (Gen 15:19).

This was the Greek tribe of Kadmos of Tyre who were indeed displaced during the Israelites' conquest of Canaan and returned to Greece bearing the alphabet, the "*writing of God*" with them. This "***Writing of God***" (Ex 32:16), the first alphabet, was written by the finger of Yehovah and given to Moses on tablets of stone (Ex 31:18). As Greek legend informs us, Kadmos (or Cadmus) returned from Canaan to found the city of Thebes and usher Greece into the age of alphabetic literacy - the most important innovation in human history!

By virtue of this power, the Greeks evolved from pirates and mercenaries into intellectual giants who created the model of western science and civilization. The Hebraic tradition and the Greek (that is, *Western*) tradition have been locked in mortal combat for cultural supremacy ever since, causing spiritual warfare in the Christian Church from its origin unto the present day. This book is not only about that spiritual struggle but also the untold story of the survival of ***The Hebrew Gospels*** and the underground Messianic Church throughout the ages.

## Reader Reviews of Sons of Zion vs Sons of Greece
(Taken from reader reviews in Amazon Books)

My mind has been continuously blown/thrilled by what I've read and I have no doubts it will be blown many more times before this final chapter is complete! There is such a beautiful mixture of historical/scientific fact and incredible spiritual revelation in this book! Yehovah bless you for writing this and sharing this information, and bless the writing of Volume 2! Knowing the history of the early church, the history of the Messianic believers is something everyone who follows Yeshua needs to know! - **Logan**

I purchased this book and left a five-star review! I'm 2/3s of the way through it and found it and to put down. The review was tough to do as there is so much good info, it was hard to describe to the public. A great *"Hebrew Root Primer"* or a *"Messianic Church Foundational Basics."* I cannot imagine the hours of research & effort you put into this book. I can say well done! - **Eric**

Truth is being made flesh! You are giving me a greater understanding of my roots, a case of lost or mistaken identity. I am excited & can hardly wait for your second volume! - **Douglas**

Impressive and thought provoking - a confluence of significant biblical issues. Your work and investment to produce this is appreciated. He is Risen! - **Ed**

I have found your new book extraordinary, amazing & thoroughly scholarly. It *"has blown me away"*!!! - **Paul**

The author, by the Grace of God, has produced a valuable work for believers and non believers alike in those end times and has revealed a rare commodity by today's standards, THE TRUTH. Do not thik that this book is a boring, dry, "Religious" dissertation—it is not. It is a fascinating read! - **Arik**

You know how a veil can drop, and one can see something in a whole new way, while wondering how it had been there all along and be unseen—the way you explained the reality and impact of the alphabet being given to humanity at Horeb. - **Henry**

Raises the standard for research! - **Edward**

# Messianic Church Arising!

### The Greatest Story Never Told – continues in Volume Two

*The Hebrew Gospels* are the most forbidden books in the history of humankind! They were targeted for extinction along with the Messianic Church itself –which was the original church of Yeshua the Messiah, better known as Jesus Christ. Both the Messianic Church and *The Hebrew Gospels* survived and spread! They appeared again in Medieval Spain. Jews by the hundreds of thousands had converted to belief in the Messiah, many were forced but far more converted of their own free will. Many believed the Hebrew story of the Hebrew Messiah belonged in the holy tongue – Hebrew – and they were restored from surviving manuscripts. The Spanish Inquisition was implemented to crush this resurgent Messianic Church!

All over Europe and elsewhere the Messianic Church had spread. Pursued by the Inquisition, these Neo-Messianics became the Underground Church and went by many names. There were differences in doctrine but always and everywhere the goal was to restore the original Messianic-Apostolic Church of the first century. They preserved *The Received Text* of the Bible, which came from the Hebrew, rather than the Greek, including *The Hebrew Gospels*. Many thousands of Messianic believers were tortured and slain. The Neo-Messianic movement of the Middle Ages preserved *The Received Text* which was the basis of the Reformation Bibles. Within thirty years of its publication – two thirds of Europe became Protestant! Messianics fled the Inquisition to the **New World. They became masters of world trade and created a surrogate pirate army and fleet – the dreaded Buccaneers of the Caribbean! They were financed and led by Messianics who brought the Spanish Empire to its knees, helped liberate the emerging countries of the New World, and snuffed out the holy terror of the Inquisition!**

First Page of Messianic Church Arising! Vol. 2 of ***Sons***

# Chapter One
## The New Jerusalem

***The captivity of Jerusalem, which is in Sepharad, shall possess the cities of the south.*** (Obadiah 1:20)

Rome fell in 410 AD to Alaric's Visigoths. The Classical Age had come to an end. "*As the Western Empire died, it left behind it empty cities with marble ruins lying like great skeletons at their centres.*"[1] The pagan Greeks (and Romans) had many faults but a lack of curiosity of the mind was not one of them. Despite their superstition, the light of reason shone brightly, creating the Western model of science and civilization - but their white marble cities are now skeletons. The Dark Ages have begun.

The Roman Church had many good qualities, but freedom of thought was not one of them. Scripture in Greek and Hebrew were banned. As a result, the study of those languages withered away. Many priceless classics of the Greeks were lost. For a millennium, the rare scholar who studied Hebrew hastened to declare his intent was solely to expose the lies and perfidy of the Jews. Simply to challenge the dogma of the Church was to court death.

The light came from the East - and it came through Sepharad. The Hebrew word '*Sepharad*' means '*land of the Book.*' The Sephardim were, therefore, the '*people of the Book.*' It is true that the diaspora Jews were more Hellenized, but it was those diaspora Jews who would keep both Hebrew language and Greek literature alive for the next thousand years - along with Hebrew Scriptures - including ***The Hebrew Gospels***!

In 1971, excavators of the Church of the Holy Sepulcher in Jerusalem broke through a wall and discovered a room dated to the early second century - when the Messianic Church was still predominate in Israel - but was being forced to flee! Excavators discovered a "*red and black drawing of a... sailing vessel with the inscription 'Domine Ivimus - Latin for 'Lord, we went'.*" Where did they go? They went across the sea to Sepharad (Spain) - '*the New Jerusalem*' - where the largest and most prosperous community of Jews, and Messianics, was a refuge for Hebrew emigrants from a century and a half of Roman war against Israel.

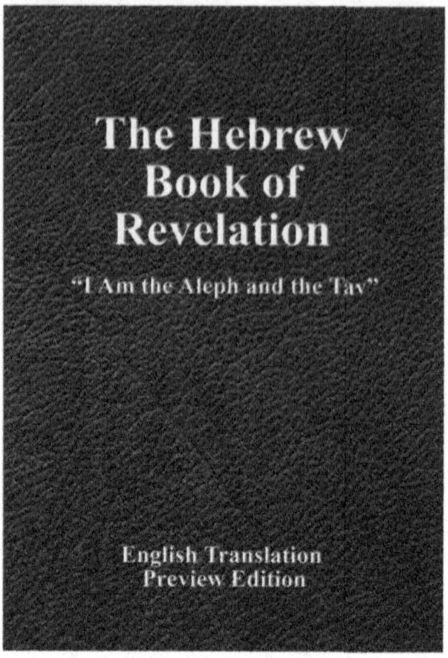


# The Hebrew Book of Revelation

## "I am the Aleph and the Tav"

Miles R. Jones, Ph.D.

(Hard Cover Preview Edition)

"*I am the Alpha and the Omega*" is what we read in English texts taken from the Greek. It seldom occurs to readers that many of the books in the New Testament were originally written in Hebrew - long before Greek-speaking believers became so numerous they dominated the Church. Therein lies the story of this book and what it reveals to us. This project is an outgrowth of the amazing discovery in 2014 by Dr. Miles R. Jones of the oldest surviving Hebrew manuscript of the New Testament Gospels, Vatican ebr. 100, (called the Hebrew Gospels from Catalonia or HGC) found bound behind the tales of Sinbad in Hebrew. As he researched the provenance of this document, and began to translate it, Dr. Jones realized that the details in this document were quite different from those of the standard Bible of today which comes from the Greek.

He painstakingly documented these differences, drawing from quotations of the Early Church Fathers for authentication. When he cross-referenced them to the Hebrew Gospel of Matthew by George Howard (often referred to as the Shem Tov Hebrew Matthew), he found many agreements in the two texts that were not in the Greek text.

Now we have various manuscripts including the Cambridge MS Oo.1.16.2, Gaster MS 1616, Freiberg MS HS-314, Paris MS 131 and the Hazon-Sloane MS 273. This is the first compilation ever of those Hebrew manuscripts of the Book of Revelation.

**Sign up at WritingOfGod.com**

First Page of chapter 1 of The Hebrew Book of Revelation

# Chapter 1

The Hebrew translation is given in **bold** - the King James Version (KJV) is included for comparison.

**1:1 The Revelation of Yeshua the Messiah, which Yehovah* gave to him (John - Yohanan in Hebrew), to reveal to his servants the things that they must do at the time of the harvest.**

> *Note: The abbreviation ה' (Ha-shem) is used in the Cambridge text as a substitute for the sacred name יהוה (Yehovah). The Gaster manuscript inserts אלהים (Elohim) for יהוה (Yehovah). Yehovah, with the correct vowel pointing - is used in many, if not most, Messianic manuscripts. Therefore, the correct name of God, Yehovah, will be used whenever indicated in the Cambridge, or other manuscripts.

KJV: The Revelation of Jesus Christ which God gave unto him to shew unto his servants which must shortly come to pass; and he sent and signified *it* by his angel unto his servant John.

**1:2 *Elohim* gave notice by his servant, John (Yochanan), who delivered the Word of Yehovah, along with the testimony of Yeshua the Messiah, about all that he saw.**

KJV: Who bare record of the word of God, and of the testimony of Jesus Christ, and of all things that he saw.

**1:3 Blessed is the one who hears and proclaims the words of this prophecy and keeps what is written in it - for the time is near.**

KJV: Blessed is he that readeth, and they that hear the words of this prophecy, and keep those things which are written therein: for the time is at hand.

# Appendix C

Bar Mitzvah
Accelerated Study Course

## בֵּית אֱמוּנָה Household of Faith

### B'nai Emunah Institute for Accelerated Learning

*Let us do good to all, but especially to those of the household of faith. Sefer Galatim 6:10*

**The Great Secret comes from Mount Sinai where believers were admonished to learn the Word and the Writing of God and teach them to their children.** Over the centuries, scattered all over the globe, Jews rose to positions of prominence, wealth and high achievement in every state, in every profession. In modern times Jews have received 20% of all Nobel Prizes ever granted despite being a fraction of a percent of world population. In the U.S. where Jews are 1.7% of the population, 40% of all Noble Prizes given to Americans have gone to Jewish-Americans. The good news is that the great secret of their success, the call from Sinai, is for all believers!

**Benai Emunah/Jones Geniuses programs focus on mastering the great secret to help students and parents achieve phenomenal results** in mathematics, reading, language, and history. Our unique accelerated programs catapult children's learning and take advantage of their marvelous God-given potential. Each program is designed so that you and your children can study at home on your own, or with support online. Our programs are different than traditional learning. We focus on building a strong foundation of basic skills, and then we teach students rapid mental mathematics, speed reading skills and memory techniques. These skills allow our students to progress much, much faster than with a traditional curriculum. By the end of first semester mathematics, most students are able to do math calculations mentally faster than an electronic calculator! No other program teaches children these important mental math, speed reading, and memory skills.

**For more information go to jonesgeniuses.com**

## THE BAR MITZVAH PROGRAM THAT COMES WITH AMAZING RESULTS OF ACCELERATED LEARNING

Believers were admonished from Sinai to master the reading of Hebrew Scripture. That involved numbers and mathematics as well, since the letters were also used as numbers. In truth, Scripture provides us with the keys and the mandate to unlock the entire mechanism of God's universe - math, science, languages & literature. B'nai Emunah programs are different than traditional learning programs. We teach the Word and the Writing of God in Hebrew, and focus on building a strong foundation of basic skills in math and literacy, then we teach students, of all ages, rapid mental math, speed reading skills and memory techniques. These skills allow our students to progress much faster than with a traditional curriculum. By the end of the first semester of math - students are able to do many advanced math calculations faster than a calculator! No other program teaches students this critical mental math, speed reading, and memory skills.

**Do Bar & Bat Mitzvah with Benai Emunah!**

## LOST HEBREW GOSPELS DECIPHERED

**First page of ancient text with title overlaid**

In the turbulent times of Medieval Spain a community of converted Jews dared to restore the Messianic Church of the first century - when Yeshua Ha Mashiach and his Apostles still lived. How do we know this? We have their *Hebrew Gospels*! The Spanish Inquisition was begun in 1478 specifically to crush the Messianic Church in Spain. In 1492, King Ferdinand & Queen Isabella completed the Reconquista, removing the last stronghold of Islam at Granada. Immediately they turned to the Jews to expel them from Spain and seize their assets. On the deadline of August 31, 1492 - Columbus' three ships left Spain for the New World filled with Jews fleeing the Edict. Messianics fleeing the Inquisition took *The Hebrew Gospels from Catalonia* with them. They ended up in the Vatican Library bound beneath another Hebrew manuscript, undiscovered for 500 years. Dr. Jones is now translating these lost *Hebrew Gospels*.

# All New *Writing of God* Video Course

Dr. Jones' *Writing of God* 18-study video course is an examination of the geographical, archaeological and scientific evidence of the Exodus. He walks you through each step to the real Mount Sinai in Saudi Arabia and much more. Learn the origins of the alphabet and how to discuss this evidence with friends and colleagues. This course of study is for those who want to the truth of the Biblical accounts. This is perfect for Bible Studies or home school education in ancient history and apologetics.

Included with the fully produced 18 video lessons is a fully illustrated *Personal Study Journal* and transcript along with the *Writing of God* text book. Optionally, you can also earn three (3) college credit hours upon completion or the course and final exaime. An online version is also availble.

## Sign up at WritingOfGod.com

# Hebrew Gospel Video Course

Dr. Miles Jones has uncovered the earliest Hebrew manuscripts of the Gospels and authenticated their first century origins!

The translation of these Gospels reveal a new depth of knowledge of our Hebrew Messiah, Yeshua. In addition to the Gospels, new manuscript discoveries that include the entire Hebrew New Testament are now being translated.

This 10-part video course explores the "hidden mana" that these Hebrew manucripts reveal in the text. The course explores the history of how these manuscripts were created and recovered and bring a significant context and insight to the biblical texts.

Dr. Jones has recorded these 10 video lessons and each lesson has been captured in the Personal Study Journal that includes graphics and picture and a multiple choice quiz and essay questions. Students can register to earn an optional two college credit hours for a minimal fee.

## Sign up at WritingOfGod.com

# The Call from Sinai - Ten Keys Scriptural Basis of Home Schooling

## From *The Writing of God* by Dr. Miles Jones

1. The call from Sinai is a sacred Covenant which God intended to reach the whole world... *"Now therefore, if you will obey my voice indeed, and keep my covenant then you shall be a peculiar treasure unto me above all people, for all the earth is mine. And you shall be unto me a kingdom of priests..."* (Ex 19:5-6)

2. It is written by God's own hand; it could be important... *"And He gave unto Moses, when he had made an end of communing with him upon Mount Sinai, two tablets of testimony, tablets of stone, written with the finger of God."* (Ex 31:18)

3. The Sinai Covenant is a teaching Covenant... *"And the LORD said unto Moses, come up to me into the mount, and be there: And I will give you tablets of stone, and a law, and commandments which I have written; that you may teach them."* (Ex 24:12)

4. God stipulates two purposes of the Covenant, His word and His writing... *"And the LORD said unto Moses, write these words, for after the purpose and character of these words I have made a covenant with you and with Israel"* (Ex 34:27 Amplified Bible).

5. God provided the Israelites with a system of writing... *"And Moses turned, and went down the mount, and the two tablets of the testimony were in his hand: the tablets were written on both their sides; on the one side and on the other were they written. And the tablets were the work of God, and the writing was the writing of God..."* (Ex 32:15-16) God specified the writing of God, the alphabet, is of Him. *"I am Alpha and Omega, the beginning and the ending... the first and the last: what you see, write in a book..."* (Rev 1:8-11). The Covenant included numeracy - letters were also used for numbers. *"Take you the sum...and divide..."* and multiply & do fractions & measurement, etc. (Num 1:1).

6. The Covenant bestows a blessing on those who answer the call from Sinai… "*And He has filled him with the spirit of God, in wisdom, in understanding, and in knowledge...*" (Ex 35:31) "*And He has put it in his heart that he may teach...*" (Ex 35:34). God specified the power of this Covenant blessing to His believers… "*Behold I make a covenant... I will do marvels, such as have not been done in all the earth... with you*" (Ex 34:10).

7. The Covenant is for men… and women… "*And every able and wise-hearted man in whose mind the Lord had put wisdom and ability, everyone whose heart stirred him up to come to do the work...*" (Ex 36:2) "*And they came, both men and women, as many as were willing-hearted…*" (Ex 35:22), and children… Daniel and his cohort were learning the word & the writing of God. "*God gave them knowledge and skill in all learning and wisdom…*" (Dan 1:17 also Ex 35:31).

8. Scripture specifies a Godly Yardstick - a measure for those receiving the blessing of the Covenant... "*In all matters of wisdom and understanding, that the king inquired of them, he found them ten times better than all the* [wise men] *that were in his realm*" (Dan 1:20).

9. The Covenant commanded believers to teach the word & the writing of God to their children... "*Said the Lord, I will put my law in their inward parts, and write it in their hearts; and will be their God, and they shall be my people*" (Jer 31:33). "*You shall read this law...*" (Deut 31:11) "*And you shall write them* [words of the law] *upon the posts of your house, and on your gates*" (Deut 6:9). "*And you shalt teach them diligently unto your children, and shall talk of them when you sit in your house, and when you walk by the way, and when you lie down,* and when you rise up" (Deut 6:7).

10. There is a curse to those who do not obey the Covenant... "*Behold, I set before you this day a blessing and a curse, a blessing if you obey…* and a curse if you will not obey…" (Deut 11:26 -28) "*My people are destroyed for lack of knowledge: because you have rejected knowledge, I will also reject you, that you shalt be no priest to me: seeing you have forgotten the law of your God, I will also forget your children*" (Hos 4:6).

# Initial Results of Jones Geniuses Accelerated Learning Program

1993-94 L.K. Hall Elementary - Out of 22 second graders, 13 read at or above 3rd grade reading level - 10 read at or above 4th grade, 7 read at or above 5th grade, 5 read at or above 6th grade and 1 read at or above 8th grade. Average - grade 3, 8th month.

1994-95 Sam Houston Elementary - First Grade, 75% were above the 50th percentile on the SABE, 25% above the 90th percentile.

1995-96 John F. Kennedy Learning Center - Sixth Grade, after only 7-10 days of Accelerated Learning, students went from 46% passing on math basic skills tests to 95% passing.

1997-99 Rogers Elementary - First Grade, 95% of students were above the 50th percentile in reading, 53% in the top quartile. In math 90% of students above 50th percentile, 56% in top quartile.

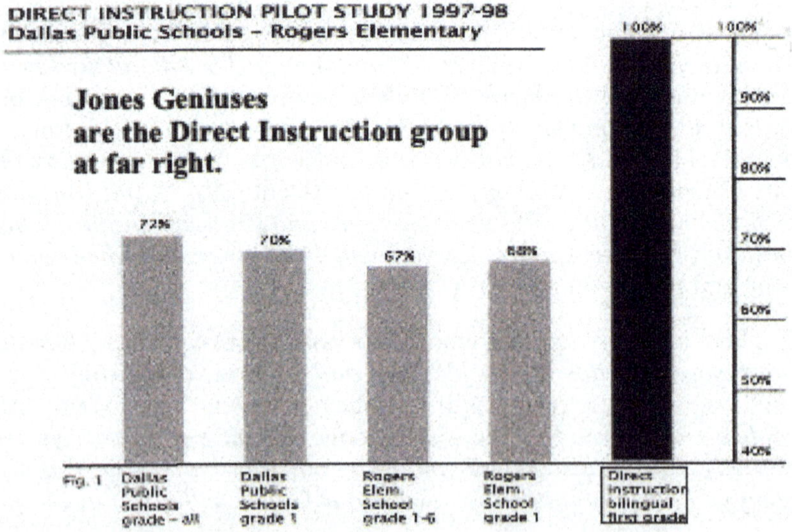


**Percent Passing Standardized SABE Reading Test**

# JONES GENIUSES IN THE NEWS

**For more information on B'nai Emunah go to jonesgeniuses.com or call 817 718 8822**

# The Mensa Contest

In 1995 inner-city Dallas 2nd graders took on MENSA PhDs in a contest of speed and accuracy calculating powers and roots. The 2nd graders did all the math mentally, the PhDs were allowed to use calculators. Teams had to answer first, and correctly to win. We spoke of the Godly Yardstick of Daniel 1:20, where the king stated Daniel and his fellows were "*ten times better than all the wise men in his kingdom*." You be the judge whether Mensa PhDs are some of the most knowledgeable men in the kingdom - and whether these children were "*ten times better!*"

## Jones Geniuses Beat Mensa PhDs 14 to 6!

**Some of the contest problems!**

$$107^2 = 11{,}449$$

$$1025^2 = 1{,}050{,}625$$

$$(\sqrt[3]{5832})^2 = 324$$

$$\sqrt[5]{2{,}887{,}174{,}368} = 78$$

$$10{,}035^2 = 100{,}701{,}225$$

$$(\sqrt[3]{373{,}248})^2 = 5184$$

$$1{,}000{,}455^2 = 1{,}000{,}910{,}206{,}025$$

$$\sqrt[3]{681{,}472} = 88$$

$$75{,}500^2 = 5{,}700{,}250{,}000$$

$$1{,}000{,}955{,}000^2 = 1{,}001{,}910{,}912{,}025{,}000{,}000$$

# Bibliography

**Abulfeda** short for El-Abd el-Fakír. Anzáh el-Mesálik ilá Ma'arifat el-Buldáni w' el-Mamálik (Light for the Road to a Knowledge of Cities and Countries).

**Aharoni, Yohanan.** 1976. "Rewriting Israel's Conquest". In Biblical Archeologist. May, volume 39, issue 2.

**Albright W. F.** 1949. The Archaeology of Palestine. Penguin Books, Baltimore, Md.

**Albright W. F.** 1966. The Amarna Letters from Palestine. Cambridge University Press, Cambridge, MA.

**Albright W. F.** 1969. The Proto-Sinaitic Inscriptions and their Decipherment. Cambridge Univ. Press, Cambridge, MA.

**Al-Ansary,** Abdul-Rahman & Majeed Khan et. al. 2002. Al-Bid: History and Archaeology. Saudi Arabia Ministries of Education, Antiquities & Museums.

**Amplified Bible.** 1987. Zondervan Publishing, Grand Rapids, MI.

**Anati, Emmanuel.** 1986. The Mountain of God. Rizzoli International Publications, NY, NY, from Editoriale Jaca, Milan.

**Anderson, Richard.** 1985. Becoming a Nation of Readers: The Report of the Commission on Reading. U. of Illinois, Champaign, IL.

**Bauer, Susan Wise.** 2007. The History of the Ancient World. W. W. Norton & Co., New York, NY.

**Beke, Charles.** 1878. Discoveries of Sinai in Arabia and of Midian. Trübner & Co., London, UK.

**Bietak, Manfred.** 1979. Avaris and Piramesse: Archaeological Exploration of the Eastern Nile Delta, London.

**Bietak, M.** 1989. "*The Middle Bronze Age of the Levant: A New Approach to Relative and Absolute Chronology*", in HML, Part 3, pp. 78-123, Gothenburg.

**Bimson, John.** 1978. Redating the Exodus and Conquest. Sheffield. Bimson, John. 1986. "Shoshenq and Shisshak: A Case of Mistaken Identity".

In SIS Review vol. 8:36-46.

**Bimson, John.** 1988. *"Exodus and Conquest - Myth or Reality"*. Journal of the American Chronology Forum vol. 2:27-40.

**Blum, Harold.** 1998. The Gold of Exodus. Simon & Schuster, New York, NY.

**Bodmer, Frederick.** 1968. The Loom of Language. George Allen & Unwin Ltd, London.

**Boren, Henry.** 1986. The Ancient World: An Historical Perspective. Prentice-Hall, Englewood Cliffs, NJ.

**Briquel-Chatonnet,** Françoise. Feb 2001. "La Révolution Alphabétique" in L'Ecriture: Ses Diverses Origines, in Les Dossiers d'Archéologie. Editions Faton,

**Dijon, France. Brugsch, Henry**. 1874. The Exodus and the Egyptian Monuments. In History of Egypt Under the Pharaohs Derived Entirely from the Monuments, Murray Press, London.

**Burton, Richard.** 1884. The Land of Midian Revisited. C. Kegan Paul & Co., London, UK.

**Burton, Richard.** 1879. The Land of Midian. C. Kegan Paul & Co., London, UK.

**Burton, Richard.** 1878. The Gold Mines of Midian. Kegan Paul & Co., London, UK.

**Caiger, Stephen.** 1936. Bible and Spade. Oxford Univ. Press, London.

**Caldwell, Penny.** 2008. God of the Mountain. Bridge Logos Publisher, Alachua, Florida.

**Carroll, James** 2001. Constantine's Sword. Houghton Miflin, NY.

**Ceram, C. W.,** aka Kurt W. Marek. 1955. The Secret of the Hittites. Phoenix Press, London, UK, reprinted from 1956 edition by Alfred A. Knopf, NY, NY.

**Ceram, C. W.** 1967. Gods, Graves and Scholars. Random House, NJ, reprinted from 1951 edition by Alfred A. Knopf, New York, NY.

**Clark, Arthur.** May-June, 1989. Lakes of the Rub al Khali. Aramco World magazine.

**Cornuke, Robert** 2000. In Search of the Mountain of God: The Discovery of the Real Mount Sinai. Broadman & Holman Publishers, Nashville, TN.

**Cornuke, Robert.** 2005. Relic Quest: Legend Chaser. Tyndale House Publishers, Carol Stream, IL.

**Cross, Frank.** 1962:225-259. "Yahweh and the God of the Patriarchs", in Harvard Theological Review, vol. 15.

**Coulmas, Florian.** The Blackwell Encyclopedia of Writing Systems. Blackwell Publishers, Oxford, UK.

**Courville, Donovan.** 1971. The Exodus Problem & its Ramifications. Crest Challenge Books. Loma Linda, CA.

**Daniels, Peter & William Bright**. 1996. World's Writing Systems. Oxford University Press, Oxford, UK.

**Darnell, John.** 2006. "Two Early Alphabetic Inscriptions from The Wadi el-Hohl: New Evidence for the Origin of the Alphabet from the western desert of Egypt", from Annual of the American Schools of Oriental Research, 59/2.

**Davies, Vivian.** Interview in The Times of London, 28 July 2003.

**DeLoach, Charles.** 1995. Giants: A Reference Guide from History, the Bible, and Legend. Scarecrow Press, Meuchen, NJ.

**Dever, William.** 1993. Recent Archaeological Discoveries and Biblical Research. University of Washington Press.

**Dever, William.** 2003. Who Were the Early Israelites and Where did they Come From? Eerdmans, Grand Rapids, MI.

**Diringer, David.** 1977. The History of the Alphabet. Unwin Brothers Ltd, Surrey, UK.

**Diringer, David.** 1996. The Alphabet: A Key to the History of Mankind. Manoharlal Publishers, New Delhi, India.

**Dykstra, Robert.** 1973. Teaching Children to Read.

**Fields, Weston.** 2006. The Dead Sea Scrolls. Brill, Boston.

**Fischer, Steven Roger.** 2001. A History of Writing. Reaktion, London.

**Finegan, Jack.** 1998. Handbook of Biblical Chronology. Hendrickson Publishers, Peabody, MA, first edition 1964, Princeton University Press.

**Fox, Robin Lane.** 1999. Pagans and Christians. Knopf Press, New York.

**Finkelstein & Silberman.** 2001. The Bible Unearthed: Archaeology's New Vision of Ancient Israel and its Sacred Texts. New York.

**Franz, Gordon.** 2000. "Is Mount Sinai in Arabia?" in Bible and Spade 13/4:101-113.

**Franz, Gordon.** "Mount Sinai is NOT Jebel al-Lawz in Arabia" in Bible and Spade.

**Franz, Gordon.** Oct 2009. "Yahweh Inscription Discovered at Mount Sinai", Associates for Biblical Research.

**Fritz, Glen.** 2006. The Lost Sea of the Exodus. Instantpublisher, US.

**Galil, Gershon.** 2010. "Most Ancient Biblical Inscription Deciphered". University of Haifa.

**Garbini, Giovanni.** 1979. Storia e Problemi dell'Epigrafia Semitica (History and Problems of Semitic Epigraphy). Oriental Institute of Naples, Italy.

**Gardiner, Alan** 1955. The Inscriptions of Sinai, edited and completed by Jaroslav Cerný, Egypt Exploration Society, London, UK.

**Gates, Arthur & Guy Bond.** Oct 1936. "Failure in Reading and Social Adjustment" NEA.

**Gaur, Albertine.** 1992. A History of Writing. Cross Rivers Press, New York, first published 1987, British Library, London, UK.

**Gelb, I. J.** 1952. A Study of Writing. University of Chicago Press, IL.

**Goldwasser, Orly.** 2010. "How the Alphabet was Born from Hieroglyphs". Biblical Archaeology Review, 36:2, Mar/Apr.

**Goody, Jack** 1987. The Interface between the Written and the Oral. Cambridge U. Press.

**Gordon, Cyrus.** 1955. Homer and the Bible. Hebrew Union College Annual, vol. 26.

**Graves, Robert** 1960. Reprinted in two volumes, The Greek Myths I & II. First Folio Society: Butler & Tanner Ltd., U.K.

**Grimal, Nicolas.** 1992. A History of Ancient Egypt. Blackwell, Oxford.

**Grimme, Hubert,** 1896. Grundzüge der Hebraeischen: Akzent und Vokallehre - mit Anhang - Ueber die Form des Namens Jahwae (Guidelines of the Hebrew: accent and vocal teaching - with Appendix - Concerning the form of the name Yahweh). Freiburg Universitaes Buchhandlung, Freiburg, Germany.

**Grimme, Hubert.** 1923. Althebräische Inschriften vom Sinai (Ancient Hebrew Inscriptions in Sinai). Orient Buchhandlung Heinz Lafaire, Hannover, Germany.

**Havelock, Eric A.** 1982. The Literate Revolution in Greece and Its Cultural Consequences. Princeton Univ. Press, Princeton, NJ.

**Healy, John.** 1990. The Early Alphabet. U. California Press, Berkeley.

**Hemphill, Ken.** 2001. The Names of God. Broadman & Holman, TN.

**Herm, Gerhard.** 1977. The Celts: The People Who Came Out of the Darkness. St. Martin's Press, New York, NY.

**Herodotus.** The Histories, 5th century BC.

**Herzog, Zeev.** "Deconstructing Walls of Jericho". Ha'Aretz, 10/29/1999.

**Hitti, P. K.** 1961. The Near East in History: A 5000 Year Story. Van Nostrand Co, Princeton, NJ.

**Hoffmeier, James K.** 2006. Israel in Egypt. Oxford University Press.

**Hoffmeier, James K.** Jan-Feb,2007:30-41. *"The Archaeological Context of the Exodus: Out of Egypt."* Biblical Archaeology Review.

**Hoffmeier, James K.** and Alan Millard, eds. 2004. The Future of Biblical Archaeology. Eerdmans, Grand Rapids, MI.

**Homer** (circa 750 BC). The Iliad. Translated by Robert Fitzgerald. Oxford University Press, Oxford, 1998.

**Jean, Georges.** 1992. Writing: The Story of Alphabets & Scripts. English translation by Harry Abrams, Inc., New York, from original French by Gallimard 1987.

**Jones, Floyd.** 1993. The Chronology of the Old Testament. Reprinted in 2005 by Master Books, Green Forest, AR.

**Jones, Miles R.** 2013. In Faith for All of Life. Publication of the Chalcedon Foundation; *"Evidence of the Exodus"* March/April pp. 4-10. *"The Stones Will Cry Out,"* May/June issue, pp. 4-11. *"The Call from Sinai,"* July/August issue, pp. 12-17.

**Josephus, Flavius** (1st century AD) Antiquities of the Jews. 1998 translation by William Whiston. T. Nelson Publishers, Nashville, TN.

**Kenyon, Kathleen.** 1957. Digging Up Jericho. Praeger Publishers, N.Y.

**Kerkeslager, Allen.** 1998. *"Jewish Pilgrimage and Jewish Identity in Hellenistic and Early Roman Egypt,"* pp. 99-225 in Pilgrimage and Holy Space in Late Antique Egypt, edited by David Frankfurter, Religions in the Graeco-Roman World, 134: Brill, Leiden, Holland.

**Keyser, John.** *"A New Look at the Dynasties of the Exodus."* Hope-of-Israel.org.

**Kim, Sung Hak.** 2006. The Burning Bush. (in Korean) tptpress@duranno.co.kr. ISBN 978-89-531-0839-4.

**King James Bible,** 1964. Kirkbride Bible Company, London, U.K.

**Kitchen, Kenneth.** 1973. The Third Intermediate Period in Egypt. Warminster, UK.

**Kitchen, Kenneth.** 1995. Biblical Archaeology Review, March-April, p.52.

**Kitchen, Kenneth.** 2003. On the Reliability of the Old Testament. William B. Eerdmans Publishing, Cambridge, U.K.

**Lappin, David.** *'The Decline and Fall of Sothic Dating"* in the Journal of the Ancient Chronology Forum, 9 (2002:83).

**LeMaire, André.** 1957. Naissance du Monothéisme. Bayard, Paris, Fr.

**Lemche, N.P.** 1985. Early Israel. E.J. Brill, Leiden.

**Logan, Robert K.** 1986. The Alphabet Effect. William Morrow, NY.

**Lyle, Anthony.** 2012. Ancient History: A Revised Chronology. Author House, Bloomington, IN.

**Mazar, Benjamin.** 1986. The Early Biblical Period, Historical Studies, edited by Ahituv & Levine, English translation. Jerusalem.

**Mahoney, Tim.** 2015. Patterns of Evidence: The Exodus (DVD). Mahoney Media, Minneapolis, MN.

**McCarter, P. Kyle.** 1996. Ancient Inscriptions: Voices from the Biblical World. Biblical Archaeology Society, Washington, DC.

**McLuhan, Marshall & Robert Logan.** Alphabet Mother of Invention. Et Cetera, vol 34, Dec, 1977.

**Merrill, Eugene.** 1996. Kingdom of Priests. Baker Books, Grand Rapids, MI.

**Millard, A. R.** 1983. Essays on the Patriarchal Narratives. Eisenbrauns Publishers, Winona Lake, IN.

**Möller, Lennart.** 2008. The Exodus Case. Scandinavia Publishing House, Copenhagen, Denmark.

**Moore, Hyatt,** ed. 1991. The Alphabet Makers. SIL Summer Institute of Linguistics, Huntington Beach, CA in conjunction with the Museum of the Alphabet, Waxhaw, NC.

**Moziani, Eliyahu** (aka Dr. Ewald Metzler). 1984. Torah of the Alphabet. Baalschem Press, Herborn, Germany.

**Musil, Alois** 1926. The Northern Hejaz. American Geographical Society, New York, NY.

**Naveh, Joseph,** 1982, Early History of the Alphabet, Magnes Press, Hebrew University, Jerusalem.

**Newton, Isaac.** 1728. The Chronology of Ancient Kingdoms Amended. Reprinted by Master Books, Green Forest, AR.

**Needham, Joseph.** 1998. Language and Logic in Traditional China. Cambridge University Press, MA.

**Norvill, Roy.** 1979. Giants: The Vanished Race of Mighty Men. The Aquarian Press, Wellingborough, Northhamptonshire, UK.

**Olbermann, Julian.** 1938. The Archaic Inscriptions from Lachish: A Non-Phoenician System of the North Semitic Alphabet. American Oriental Society, Baltimore, MD.

**Parkinson, Richard** 1999. Cracking Codes. Univ. of California Press, Berkeley, CA.

**Petrie, Flinders.** 1891. Researches in Sinai. John Murray, London, UK.

**Petrovich, Douglas.** 2016. Worlds Oldest Alphabet. Carta, Jerusalem.

**Philby, Henry Saint John**. 1957. The Land of Midian. Ernest Benn Limited, London, UK

**Quayle, Stephen.** 2005. Giants. Thunder Publishers, Bozeman, MT. Reader's Hebrew Bible. 2008. Zondervan, Grand Rapids, MI.

**Reeves, Nicholas & Richard Wilkinson.** 1996. The Complete Valley of the Kings. Thames & Hudson, London, UK.

**Robinson, Andrew.** 1995. The Story of Writing. Thames & Hudson, UK.

**Rohl, David.** 1995. Pharaohs and Kings. Crown Publishers, New York, NY. Originally published in the UK as A Test of Time. Rohl, David. 2003. From Eden to Exile: The 5000 Year History of the People of the Book. Greenleaf, TN.

**Rohl, David.** 2015. Exodus: Myth or History? Thinking Man Media, St. Louis Park, MN.

**Salibi, Kamal** 1985. The Bible Came from Arabia. Jonathan Cape, London, UK, first edition, Die Bibel kam aus dem Lande Asir, Rowohlt Verlag, Hamburg, Germany.

**Sandars, N. K.** 1978. The Sea Peoples: Warriors of the Ancient Mediterranean. Thames & Hudson, London.

**Schlain, Leonard** 1998. The Alphabet versus the Goddess: Conflict between Word and Image. Penguin, NY.

**Sass, Benjamin,** 1988. The Genesis of the Alphabet and its Development, in 2nd Millenium B.C. Agypten und Altes Testament 13, Wiesbaden.

**Sass, Benjamin.** 1991a. Studia Alphabetica: On the Origin and Early History of the Northwest Semitic, South Semitic and Greek Alphabets. Universitätsverlag Freiburg. Schweiz, Vandenhoeck & Ruprecht Göttingen, Germany.

**Sass, Benjamin.** 1991b. *"Beth Shemesh Tablet and the Early History of the Proto-Canaanite Cuneiform and So. Semitic Alphabets."* UF 23:3-15-326.

**Sass, Benjamin.** 2005. Alphabet at the Turn of the Millenium. Tel Aviv.

**Schmandt-Besserat, Denise.** 1996. How Writing Came About. The University of Texas Press, Austin, TX.

**Smelik, Klaas.** 1991. Writings from Ancient Israel. Westminster John Knox Press, Louisville, KY.

**Socrates** (circa 469-399 BC). Quoted in The Dialogs of Plato, also in Plato: The Last Days of Socrates. 1993. Translated by Hugh Tredennick and Harold Tarrant. Penguin Books, London.

**Strong, James.** 1970. Exhaustive Concordance of the Bible. Hendrickson Press, Peabody, MA.

**Thomas, Carol & D. P. Wick.** 1994. Decoding Ancient History. Prentice Hall, Englewood Cliffs, NJ.

**Ullman, B. L.** Jul-Sept, 1927:311-328. *"The Origin and Development the Alphabet"*, from American Journal of Archaeology, vol. 31/3. Ussher, James. 1658. The Annals of the World. 2005 reprinted by Master Books, Green Forest, AR.

**Wallenfels, R.** 1983 & 1986. Redating the Byblian Inscriptions. JANES 15:79-113.

**Williams, Larry.** 1990. Mountain of Moses. Wynwood Press, NY, NY.

**Wilson, Marvin** 2001. YHWH in the Bible. Pilgrim Publishing Co., TX.

**Winnett, F. V. & W. L. Reed.** 1970. Ancient Records from North Arabia. University of Toronto Press.

**Wood, Bryant.** 1990. *"Dating Jericho's Destruction"*, Biblical Archaeology Review, 5.

**Wood, Bryant.** 2002. *"The Bible versus Modern Scholarship",* A Good News Interview by John Elliot. United Church of God.

**Wood, Bryant.** 2008. *"Did the Israelites Conquer Jericho"*, Associates for Biblical Research, May 1, 2008.

**Wyatt, Ron.** 1998. Discovered - The Exodus (DVD). Wyatt Archaeological Research, Cornersville, TN.

**Yamauchi, Edwin.** March-April, 2007:29-37. *"Historic Homer: Did it Happen?"* in Biblical Archaeology Review.

**Yardani, Ada.** 2014. Understanding the Alphabet of the Dead Sea Scrolls. Carta, Jerusalem.

**Zali, Anne & Annie Berthier.** 1997. La Naissance des Ecritures. Bibliothéque National de France, Paris.

Made in United States
North Haven, CT
10 November 2026

22084528R00124